NATURAL
HISTORY
MUSEUM

DARWIN

THE MAN, HIS GREAT VOYAGE, AND HIS THEORY OF EVOLUTION

JOHN VAN WYHE

ANDRE
DEUTSCH

THIS IS AN ANDRE DEUTSCH BOOK

Design copyright © André Deutsch Limited 2018

Text copyright © John van Wyhe 2008

First published in 2018 by André Deutsch, a division of the Carlton
Publishing Group, 20 Mortimer Street, London, W1T 3JW

The content of this book first appeared in *Darwin*, ISBN
978-0-233-00251-4, André Deutsch, 2008.

The right of John van Wyhe to be identified as the
Proprietor of this work has been asserted in accordance
with the Copyright, Designs and Patents Act 1988.

Printed in Dubai

A CIP catalogue for this book is available from the British Library

ISBN: 978-0-233-00536-2

CONTENTS

INTRODUCTION

Charles Darwin forever transformed our understanding of life on Earth. Darwin is remembered primarily for his theory of evolution by natural selection – a great synthesis that has become the unifying bedrock of the life sciences.

Darwin was the first to be able to explain, by purely natural causes, where the different kinds of living things came from and how they become exquisitely adapted to their particular environments. The true story of how evolution by natural selection was uncovered is quite different from that familiar to most people. Darwin did not challenge a world full of young earth creationists. Even today the most dramatic evidence for evolution are those broad features about geology and biodiversity first uncovered by mostly Christian naturalists before Darwin. By looking at these discoveries as they unfolded, it is possible to make evolution, and Darwin's breakthrough, more intelligible. Before Darwin sailed on the *Beagle* it was almost universally accepted by western naturalists that the world was many millions of years old and that countless eras of life had come and gone in succession. Each era had had its own characteristic forms of life. When a species vanished from the fossil record it never returned again. It was believed that new species were somehow created to suit the new environments in new eras. It was even known that the fossil record was

progressive – in the oldest rocks there were shells, then fish, then amphibians, then reptiles and finally mammals. Another major addition to scientific knowledge was that there were a lot more species in the world than previously known. Sometimes fossils seemed to fit in the gaps between living groups. It was found that whole classes of organisms were closely related to one another through chains of similarity. Even more fundamental, it was found that all groups fit as sub-groups within larger ones. Across entire classes of organisms, the same structure appeared to have been based on the same pattern. Similarly, the study of embryos of many species showed striking resemblances during their early development, which were progressively lost as they aged. Living things were also found to have much in common in their chemical composition, while microscopes showed that all living things are made of cells.

Darwin found the naturalistic explanation for the origin of species, that is, where species come from. Like the tiny uplifts from earthquakes that ultimately pushed up the Andes, minute actions were the actual causes that had led to the massive changes of life over time as seen in the fossil record. His unmatched vision allowed him to appreciate simultaneously the minutely small and the almost impossibly vast.

Darwin showed that not only are we ourselves descended from earlier kinds of animals, but that all creatures living and extinct that have ever lived on this Earth were all produced by the same simple processes of reproduction with successful or unsuccessful descendants.

It is unfortunate that today so many people do not understand Darwin's work. Explaining the complexities of genes and DNA is not necessary to understand the basic fact of evolution. Tiny natural differences between individual organisms, such as we see between ourselves or our pets, and the fact that offspring resemble parents inexactly, is practically all one needs to understand evolution. The next step is to be able to appreciate how these simple and immediately observable processes, if reiterated long enough, accumulate to produce almost infinite change.

This is not the sort of book normally written by historians of science. And it is not written for historians. This book is an experiment in communicating the history of science, Darwin and evolution to a wider audience.

With its reproductions of original documents and hundreds of contemporary illustrations, this book will help to bring Darwin, the places he visited and the natural phenomena he studied, back to life.

THE ABYSS OF TIME

Human beings may always have wondered about the age of the Earth. In 1620, the cleric and scholar James Ussher, using sophisticated biblical scholarship for his time, gave the age of the Earth as about 6,000 years, as he believed it was created in 4004 BC.

Over the next hundred years, scholars studying the Earth itself generated a mountain of learning, discoveries and controversies. Although most of them were Christian and orthodox, from the continued investigation of the Earth's surface, they came to realize that the world was extremely ancient, in the order of millions of years old. While this was a monumental shift in perceptions of the past, scholars of the seventeenth and eighteenth centuries were not on a crusade to undermine the Bible. Instead, they were pursuing their interests in particular sciences, and sometimes devoted enormous effort to reconciling their findings with their belief in Christianity.

In the eighteenth century, the French naturalist Georges Buffon tried to estimate the age of the Earth through experiments. He prepared balls of iron in a graduated set of sizes that were heated almost to melting point and then allowed to cool in a cave. The larger the sphere, the longer it took. Buffon then calculated that it would take a ball the size of the Earth about 75,000 years to cool from a molten state.

Around the same time the geologist James Hutton believed that the Earth was a machine designed by God to sustain life. Hutton was convinced the Earth was locked in an eternal cycle of decay and restoration. Erosion wore down the land and it settled in the oceans to create strata, which were later uplifted by the Earth's internal "heat engine" to make new land.

Hutton was no armchair theorist. He tirelessly trekked across southern Scotland studying how the present surface of the Earth came to be formed. He discovered some rock formations called "non-conformities", which demonstrated that the Earth's surface had changed over long eras. For example, he found a section of strata that had been tilted until vertical, it had then been sheered off and another layer of horizontal strata

The Deluge, a mezzotint with etching by John Martin (1828) depicting Noah's flood.

THE SOLID PAVEMENT OF THE GLOBE

"We felt ourselves necessarily carried back to the time when the schistus on which we stood was yet at the bottom of the sea, and when the sandstone before us was only beginning to be deposited, in the shape of sand or mud, from the waters of a superincumbent ocean. An epocha still more remote presented itself, when even the most ancient of these rocks, instead of standing upright in vertical beds, lay in horizontal planes at the bottom of the sea, and was not yet disturbed by that immeasurable force which has burst asunder the solid pavement of the globe. ...The mind seemed to grow giddy by looking so far into the abyss of time." *The works of John Playfair.* Constable, 1822, page 80.

Opposite, top left: The French naturalist Jean-Baptiste Lamarck who proposed that living things did not become extinct but changed into higher forms.

Below: Scottish philosopher James Hutton. Etching by John Kay, 1787.

was deposited above it. Hutton showed that the rocks revealed a "succession of worlds" that had come and gone. And with regard to its age, Hutton only quipped "we find no vestige of a beginning, no prospect of an end".

Hutton's work inspired one of the most famous early nineteenth-century geologists, Charles Lyell. By this time the Bible and miracles were no longer considered acceptable explanations for geological phenomena. Too many naturalistic explanations had been found again and again to explain features of the Earth's structure. Lyell examined the Mount Etna volcano in Italy and showed that it must have grown slowly over a vast period of time. Lyell saw that events that appeared sudden on a geological scale could be the result of a long sequence of very mundane happenings.

M. le CHEVALIER de LAMARCK.
Professor of Botany of the National Institute.

This aspect of his theory is the one usually remembered as "uniformitarianism", and is held up as the key to Lyell's theories. His contemporaries, although they had no problem with mundane causes explaining past events, did object to Lyell's insistence that causes of the same intensity as those observed in their time were all that had ever existed. That meant that volcanic eruptions, earthquakes and erosion must always have occurred on the same scale.

Below: Geological unconformity on the river Jed. Plate 3 from Hutton's *Theory of the Earth*, volume 1 (1795). Tilted sedimentary beds are overlain by later horizontal beds.

THE CREATION OF NEW SPECIES

Lyell also addressed the question of the successive appearance and disappearance of fossil species in the geological record. Here too he tried to show that gradual natural processes were responsible. Because species were fixed, as the world gradually changed, species would eventually become extinct as their environments changed too much. Where did the new subsequent species come from? He accepted that species introductions were probably as piecemeal as extinctions and hypothesized that new species somehow appeared as the result of "special creations" in accordance with the new environments. Just as Lyell's *Principles of Geology* (1830–33) was coming out, a young English geologist researching in South America was able to put some of Lyell's ideas about gradualism to the test. He was also to find Lyell's dodging of the question of where new species came from not entirely satisfactory. His name was Charles Darwin.

Charles Lyell, photographed by Ernest Edwards around 1865.

CAROLI LINNÆI

EQUITIS DE STELLA POLARI,
ARCHIATRI REGII, MED. & BOTAN. PROFESS. UPSAL.;
ACAD. UPSAL. HOLMENS. PETROPOL. BEROL. IMPER.
ACAD. ~~~ MONSPEL. TOLOS. FLORENT. SOC.

SYSTEMA
NATURÆ

PER
REGNA TRIA NATURÆ,
SECUNDUM
CLASSES, ORDINES,
GENERA, SPECIES,
CUM
CHARACTERIBUS, DIFFERENTIIS,
SYNONYMIS, LOCIS.

TOMUS I.

EDITIO DECIMA, REFORMATA.

Cum Privilegio S:æ R:æ M:tis Sveciæ.

HOLMIÆ,
IMPENSIS DIRECT. LAURENTII SALVII,
1758.

THE ABUNDANCE OF NATURE

The great English naturalist John Ray had published lists of a few hundred species at the end of the seventeenth century. By the end of the eighteenth century, the number of known species had grown to hundreds of thousands.

As European ships circled the globe for the first time, new kinds of species became known, from the porpoise and the dodo to the strange marsupial creatures of Australia. Systematists, like the Swedish botanist Carl Linnaeus, created elaborate systems to arrange and sort them, and it was found that whole classes of organisms were closely related to one another through chains of similarity. Even more fundamentally, it was found that all groups fit as subgroups within larger ones, so that, for example, all species of wolves were classed with foxes, jackals and dogs as canides.

Sometimes fossils seemed to fit in the gaps between living groups. At any rate, fossils could also be classed in the same way. Across entire classes of organisms, the same structure appeared to have been based on the same pattern. Similarly, the study of embryos of many species showed striking resemblances during their early development, which were progressively lost as they aged. Living things were also found to have much in common in their chemical composition, while microscopes showed that all living things are made of cells.

Although fossils had always been around, for many centuries it was unclear whether fossils (in our sense) were anything to do with organisms. As more concerted attention was paid to these mysterious objects, it became clear that they really were the remains of living plants and animals that had become petrified, and not just sports of nature that somehow grew inside rocks. In the seventeenth century, the English polymath Robert Hooke was one of the earliest people to show that fossils were once living things, by comparing their structures with living equivalents under the newly invented microscope. The Danish physician Nicolas Steno demonstrated that so-called fossil "tongue

Opposite: Title page of Carl Linnaeus's Systema naturae (1758).

Right: The great Swedish botanist and taxonomist Linnaeus, by Magnus Hallman circa 1780.

stones" looked so much like shark's teeth because they in fact were shark's teeth turned to stone.

The brilliant French comparative anatomist Georges Cuvier, through his detailed analyses of fossil bones, first proved the fact of extinction. His excavations in the Paris basin showed that the further back in time one moved, the more different the kinds of living things. The principle was established that the deeper one went into the Earth, the more the fossils differed from modern forms, and forms that disappeared in one era never reappeared in some later era.

By the 1830s, geologists following the work of Cuvier, William Smith and William Buckland from Oxford, almost unanimously agreed that the geological record showed a progressive succession of eras of life in the history of the Earth. The earliest were shells, then came crustaceans and fish. The later rocks contained the first reptiles and the recently discovered dinosaurs, while the comparatively recent rocks contained the first mammals, although of extinct types. Even younger rocks contained fossils of extinct species similar to currently living species, and the most recent of all contained some still-living species. However, no one had ever found a fossil human, and

SÉRIE IV.

CUVIER.

Above: Engraving of the great French comparative anatomist and palaeontologist Georges Cuvier. He is depicted examining a fossilized fish with a magnifying glass.

CONSTANT AND UNCHANGEABLE

The pious John Ray compiled lists of species and insisted that "the number of true species in nature is fixed and limited and, as we may reasonably believe, constant and unchangeable from the first creation to the present day. ... A species is never born from the seed of another species". Linnaeus at one time declared "of all the species originally formed by the Deity, not one is destroyed".

English naturalist John Ray.

RECONSTRUCTING THE EARTH'S PAST

Nicolas Steno showed that a fossil must originally have been harder than the material around it, because it was the fossil that shaped the stone, thus leaving an impression, and not vice versa. So the once-soft material that came to enclose the shark's tooth must have precipitated out of suspension in a liquid such as water. This meant that the original layer was horizontal. Any deviation from horizontality indicated that the layer had later been disturbed. If there were multiple layers of rock or strata, then those underneath must be older than those deposited above them. This made it possible to distinguish different ages for rocks, and, ultimately, meant that history of the Earth's past could be reconstructed.

Nicolas Steno showed that fossils shaped the matrix around them and not the other way round, hence they were first deposited in soft material, mud.

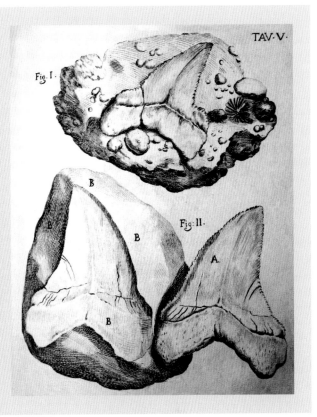

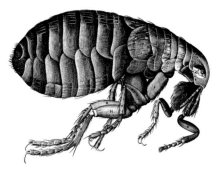

Right: Image of a flea as seen under the microscope from Robert Hooke's *Micrographia* (1664). The microscope made the majority of living things visible for the first time.

Far right: Orang-outang, female. From *Histoire Naturelle des Mammifères*, volume 3, 1819–42, by Étienne Geoffroy Saint-Hilaire and Georges Cuvier.

this made it clear that these ancient eras of life had existed before the creation of man. This led many to believe that the creation story in Genesis only referred to the most recent of many creations, the only one that concerned us.

So, by the time Darwin was a young schoolboy, most informed people knew that the world was not only 6,000 years old. The other major realization that had come about was that there were a lot more species in the world than previously imaginable.

Orang-outang, femelle

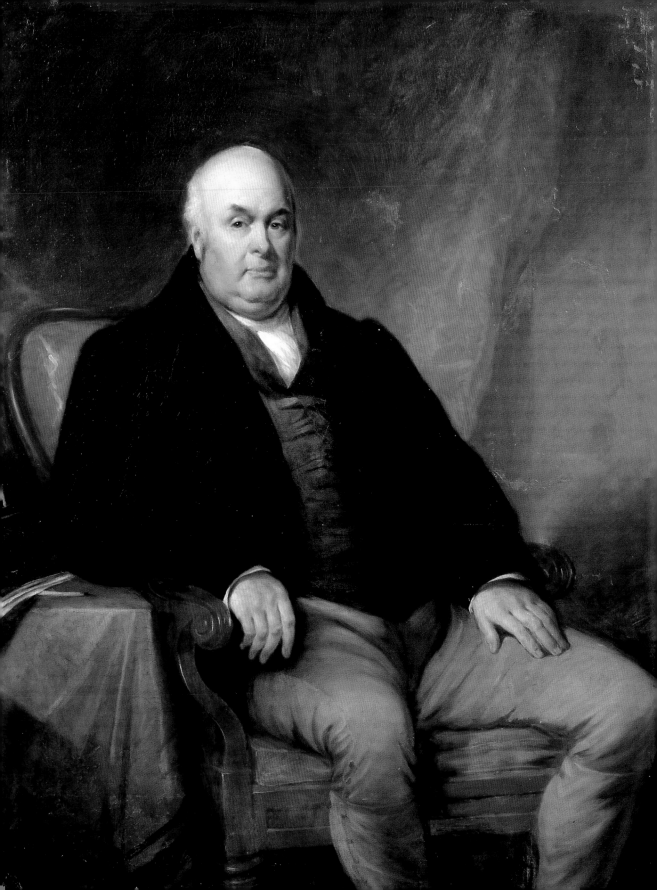

CHARLES DARWIN: BORN A NATURALIST

Charles Robert Darwin was born on 12 February 1809, the fifth of six children. He was born into a wealthy gentry family in Shrewbury, Shropshire, in the middle of Georgian England. It was rural England as evoked by Jane Austen and the Napoleonic wars seemed far away.

The family home, The Mount, was a large comfortable house with many servants. Darwin's father, the portly physician and financier Robert Darwin, was the respected son of the famous (or notorious) philosopher and poet Erasmus Darwin. Darwin's mother, Susannah Wedgwood, was the daughter of Josiah Wedgwood, the famous potter.

Charles Darwin's recollections of his earliest childhood reveal two basic and lifelong character traits: sincere modesty and an insatiable curiosity to understand

Opposite: Darwin's father Dr Robert Waring Darwin. Darwin remembered him as "the wisest man I ever knew".

Right: "The Mount", built by Darwin's father in about 1800, was Darwin's childhood home.

how things work. In later years he could only just remember the death of his mother when he was eight, and no clear evidence exists to support the frequently repeated view that her death had a profound psychological effect on him. His recollections give equal attention to his mother's "curiously constructed work-table". He was tended instead by maidservants so that the death of his mother would not have been the same sort of deprivation as it would be for a modern child. His three elder sisters oversaw the upbringing of Darwin and his younger sister Catherine.

Darwin was first tutored at home by his sister Caroline before going to a day school in Shrewsbury run by the minister of the Unitarian Chapel, which his mother attended along with the children. Unitarians formed a dissenting denomination that denied the doctrine of the Trinity. Nevertheless, Charles was baptized and meant to belong to the Church of England. This was crucial in qualifying him to attend an English University at a later date. In 1818, Darwin went to the public school in Shrewsbury, about a mile from The Mount, as a boarder, where he stayed until 1825.

Left: A rare picture of Darwin's mother Susannah Wedgwood a few years before her marriage.

Opposite, bottom: The Wedgwood family painted by George Stubbs. Darwin's mother is on horseback in the centre and his uncle Josiah or "Jos" is the next figure to the right. Darwin's maternal grandfather, the potter Josiah Wedgwood, and his wife Sarah, are seated under the tree.

CHILDHOOD MEMORIES

"Looking back as well as I can at my character during my school life, the only qualities which at this period promised well for the future, were, that I had strong and diversified tastes, much zeal for whatever interested me, and a keen pleasure in understanding any complex subject or thing. ….I remember in the early part of my school life that I often had to run very quickly to be in time, and from being a fleet runner was generally successful; but when in doubt I prayed earnestly to God to help me, and I well remember that I attributed my success to the prayers and not to my quick running, and marvelled how generally I was aided." Charles Darwin, *Autobiography,* 1958, page 43.

Darwin and his sister Catherine (1816), pastel drawing by Rolinda Sharples, about a year before the death of their mother.

Darwin was not an impressive student and he felt his time at school was wasted learning Greek and Latin classics. He studied chemistry in a home "laboratory" set up in a garden shed with his elder brother Erasmus. Together they investigated the composition of various domestic substances, by mixing, boiling, separating and crystallizing. Through these activities and the careful study of chemistry books, Darwin learned first-hand the basic principles of scientific experimentation. He later recalled:

"[The chemistry] *was the best part of my education at school, for it showed me practically the meaning of experimental science. The fact that we worked at chemistry somehow got known at school, and as it was an unprecedented fact, I was nick-named "Gas". I was also once publicly rebuked by the head-master, Dr Butler, for thus wasting my time over such useless subjects.*"

Darwin loved country sports, such as riding, shooting, fishing and solitary walks. He was not interested in social sports like cricket. Erasmus went up to Cambridge in 1822 to study medicine and then on to Edinburgh University in 1825 to continue his studies. Darwin's father thought it was a good opportunity for Charles to make a start towards the medical profession and in 1825 Darwin went to Edinburgh where a whole new world of possibilities lay before him.

CHILD COLLECTOR

"By the time I went to this school [the day school] my taste for natural history, and more especially for collecting, was well developed. I tried to make out the names of plants, and collected all sorts of things, shells, seals, franks, coins, and minerals. The passion for collecting, which leads a man to be a systematic naturalist, a virtuoso or a miser, was very strong in me, and was clearly innate, as none of my sisters or brother ever had this taste. I remember I took great delight at school in fishing for newts in the quarry pool. — I had thus young formed a strong taste for collecting, chiefly seals, franks etc. but also pebbles & minerals, — one which was given me by some boy, decided this taste. — I believe shortly after this or before I had smattered in botany."
Charles Darwin, *Autobiography*, 1895, page 22.

Shrewsbury School where Darwin lived as a border from 1818–25.

EDINBURGH UNIVERSITY

In 1825, aged only 16, Darwin joined his elder brother Erasmus at Edinburgh University to study medicine. Darwin greatly disliked his studies and was horrified at the sight of blood or operations, which were then still performed without anaesthetic.

Edinburgh University from South Bridge Street in 1829. Darwin studied medicine here for two years 1825–27.

In his second year at Edinburgh Darwin's brother left and this led Darwin to make many student friends with shared interests in natural science.

Darwin learned more about science in his own time than he learned from his lectures. In Edinburgh, Darwin first attended scientific societies and was much impressed by the world of elite men reading and debating scientific papers. He also began to read and study scientific books and journals. He was inspired to collect and investigate marine creatures in tidal pools with Dr Robert Grant, a local expert. These Darwin investigated and dissected under a "wretched microscope". "I made one interesting little discovery, and read about the beginning of the year 1826 [actually 1827], a short paper on the subject before the Plinian Soc[iet]y. This was that the so-called ova of Flustra had the power of independent movement by means of cilia,

and were in fact larvæ." Darwin at first rushed to inform Grant of the discovery, but was shocked when Grant told him that this was his area of research and that it was unfair of Darwin to publish it. So Darwin was introduced almost simultaneously to the thrill of discovering something new in nature and the scientific jealousy that often accompanies it. Afterwards, Darwin was less keen to be close to Grant.

Darwin was also first introduced to the study of geology at Edinburgh, though because of the old-fashioned views of his professor, with little positive result. As he later recalled in his autobiography:

"During my second year in Edinburgh I attended Jameson's lectures on Geology and Zoology, but they were incredibly dull. The sole effect they produced on me was the determination never as long as I lived to read a book on Geology or in any way to study the science. … I … heard Professor Jameson, in a field lecture at Salisbury Craigs, discoursing on a trap-dyke, with amygdaloidal margins and the strata indurated on each side, with volcanic rocks all around us, and say that it was a fissure filled with sediment from above, adding with a sneer that there were men who maintained that it had been injected from

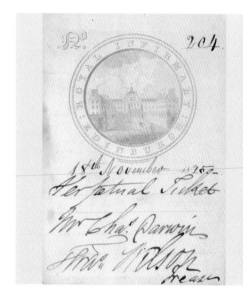

UNIVERSITY STUDIES

Darwin's father Robert, like Darwin's grandfather before him, the famous poet Erasmus Darwin, was a physician and also studied medicine in Edinburgh. It was not surprising that Darwin was first intended for the medical profession, particularly since his dislike of Latin and Greek would not allow him to pursue law, for example. Darwin paid for nine university courses over his two years in Edinburgh, covering subjects such as anatomy, surgery, *materia medica* (therapeutic substances) and the practice of physic, as well as chemistry and natural history.

An Edinburgh University ticket to gain admission to lectures dated 18 November 1825 which belonged to Charles Darwin. The printed ticket proved that someone had paid to attend.

beneath in a molten condition. When I think of this lecture, I do not wonder that I determined never to attend to Geology."

One useful skill Darwin did pick up was the art of skinning and drying birds for scientific purposes. This he learned from John Edmonstone, a freed slave and the first black man Darwin ever knew, whom he described as

"a very pleasant and intelligent man". Darwin and his family were strongly anti-slavery, and this early encounter convinced him that peoples of other races were just as human as he was.

Darwin understood that his father would leave him enough property to live in comfort, thus dispelling any real sense of urgency in learning the endless details of medicine

Below: Leith Harbour in 1825. Darwin collected sea creatures in tidal pools and hired fishermen to take him out collecting.

SEARCHING FOR SPECIMENS

In Edinburgh, Darwin's lifelong interest in marine invertebrates was kindled. He often accompanied his then scientific mentor Dr Grant to collect specimens in tidal pools along the Firth of Forth. He later recalled: "I also became friends with some of the Newhaven fishermen, and sometimes accompanied them when they trawled for oysters, and thus got many specimens." Already at this early age Darwin had begun to explore novel and unconventional methods for procuring specimens and information. He was prepared to follow any course likely to lead to new knowledge or better specimens.

Robert Edmond Grant, expert on marine invertebrates and Darwin's first scientific mentor. Grant's prickly personality and scientific jealously later led Darwin to avoid him.

necessary to become a physician. After two years it became clear to his father that Darwin did not want to be a physician, so it was proposed he become a clergyman instead. Although not particularly religious, Darwin did not doubt the truth of the Bible. Becoming a clergyman would mean he could pursue a personal fascination for natural history like the famous parson- naturalist Gilbert White. This also meant that Darwin would have to attend an English university to pursue a BA degree as the prerequisite to entering holy orders in the Church of England.

Right: English clergyman and naturalist Gilbert White whose classic work *The Natural History of Selborne* (1789) was one of the earliest scientific influences on Darwin.

Far right: Darwin's Plinian Society card, where he gave his first scientific papers on marine invertebrates.

Extract from Darwin's 1825 notes on chemistry lectures by Thomas Charles Hope. At one point he writes "I here missed two lectures on account of illness".

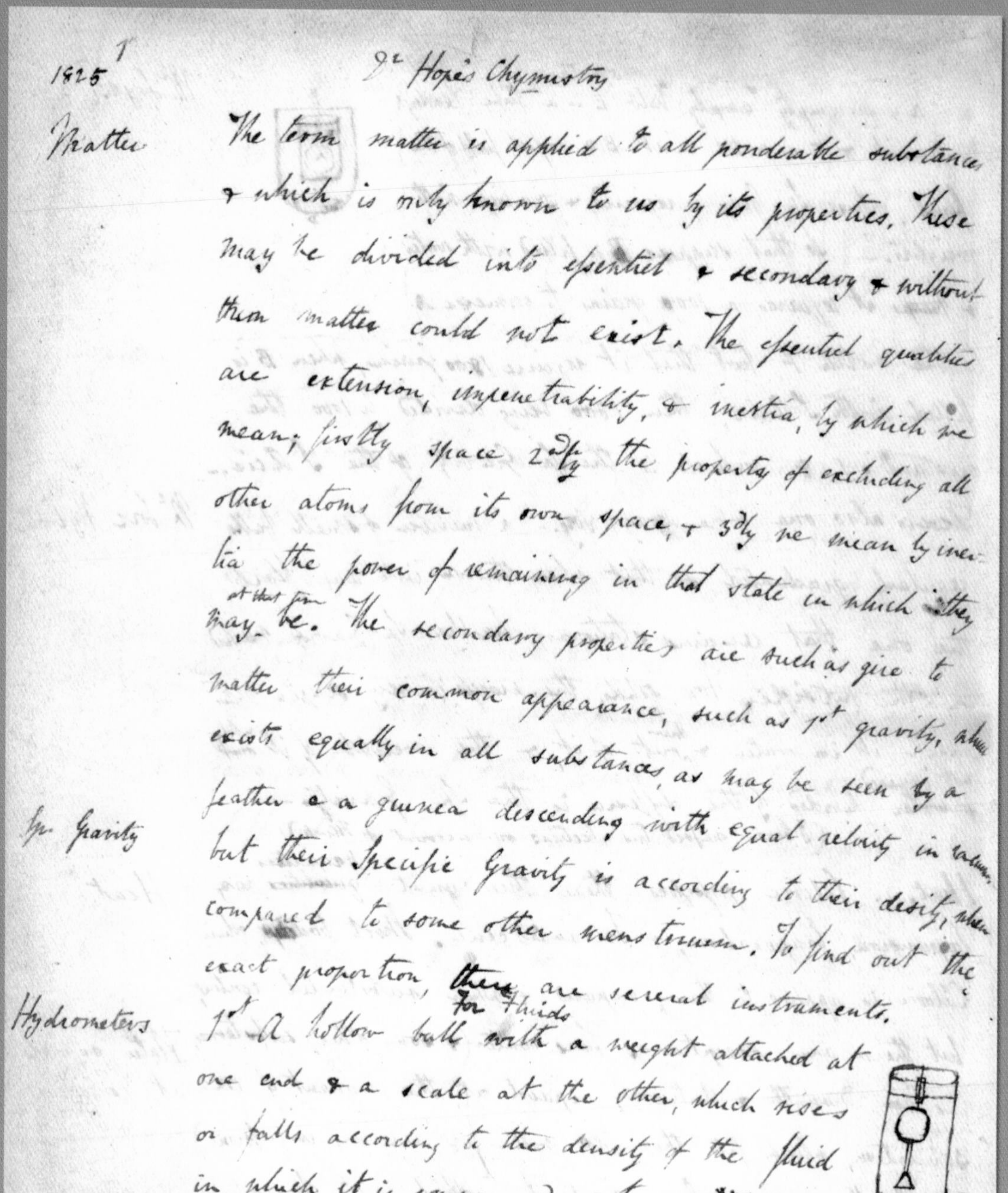

1825

Dr Hope's Chymistry

Matter

The term matter is applied to all ponderable substances & which is only known to us by its properties. These may be divided into essential & secondary & without them matter could not exist. The essential qualities are extension, impenetrability, & inertia, by which we mean, firstly space 2dly the property of excluding all other atoms from its own space, & 3dly we mean by inertia the power of remaining in that state in which they at that time may be. The secondary properties are such as give to matter their common appearance, such as 1st gravity, which exists equally in all substances, as may be seen by a feather & a guinea descending with equal velocity in vacuum

Sp Gravity

but their Specific Gravity is according to their density, when compared to some other menstruum. To find out the exact proportion, there are several instruments.

Hydrometers

1st A hollow ball for fluids with a weight attached at one end & a scale at the other, which rises or falls according to the density of the fluid in which it is immersed. to wit. —

Mr Loyds Hydrometer

2nd A is an empty empty bulb E is a tube leading
from it & F a notch in it B a vessel full of the
fluid whose Sp: Grav is required, & D a cup with
weights. — So that suppose B is filled with water
& keeps it requires a 1000 grains to immerse A
to the notch F, but that it requires 1800 grains, when B is
filled with S Acid, then 1800 being divided by 1000 the
quotient is 1.8 which is the Sp Gravity of the S Acid.—
There is also one more way, viz. — a number of small balls
regularly graduated, so that when thrown into any fluid
the one that remains stationary, has the Sp Gravity marked
on the outside. For solids, the process is very simple viz
weigh it in water & then out of it, & the weight of it when
not immersed
+ water divided by the difference is the Sp Gravity.—
(There missed two lectures on account of illness)

Heat.— Caloric possesses three three great qualities properties viz
Expansion, Evaporation, & Incandescence. Most bodies, when
Caloric is applied to them, expand & when subtracted contract,
but there are exceptions. Thus water (& some other substances,
as Iron, Bismuth & Antimony expand at the moment of con=
solidation, which fits them so well for receiving impressions)
gradually expands from 39° Far:, which is its maximum
density, to 32 when it is increased in bulk 1/10.th

Dr Hope has shown this very neatly.

$$32° + 68° \quad \diagdown$$
$$\qquad et \ vice \ versâ \qquad \Big\} \ (39½° \ is \ M.D.)$$
$$39½° + 32° \quad \diagup$$

Of this property of expansion in heat, we make various uses, of which one of the greatest is, the Thermometer. The principal fluids used in this instrument are Air, Alcohol & Mercury. Air is used for very delicate experiments. Alcohol for intense cold, & Mercury for all common purposes. The great advantage of Mercury, is that up to 212° of Far.t it expands equally, which is not the case with most other fluids. In different countries different scales are used; thus in England Farenheit, & on the Continent chiefly Celsus' or commonly called the Centigrade. from its being divided, between the freezing & boiling point into a 100:— whilst Far.t is in 180°. so that to reduce the former into the latter.

$$C° \times \frac{9}{5} + 32 = Far.t$$

In Reaumour it is

$$R° \times \frac{9}{4} + 32 = Far.t$$

There is also another kind of Thermometer, called a
Register, for which purpose they have been, various
means invented. Such as this, which is by far the
most ingenious & useful. —

A is a tube filled with Sp of Wine
from B to C is Mercury. C a float
& E a wire, faced in the float at
one end, & at the other F a
pencil. — H, a revolving cylinder covered with paper, ruled verti-
cally & horizontally, the former for the times & the
latter for the degrees. — So that the Alcohol

in A pushes up the float C. & with it the pencil
F. thus marking the cylinder, which is made to
revolve in a certain given time. —

All bodies tend to a diffusion of Caloric, yet their
capacity for it is by no means the same, thus

if equal quantities of Mercury & water be
mixed together, of different Temperatures, the
Medium will be nearest that of the water,
shewing that water has the greater capacity.

CAMBRIDGE UNIVERSITY

Darwin entered the books at Christ's College, Cambridge on 15 October 1827, but since he had forgotten much of his school Greek he had to be tutored at home before "going up" to Cambridge. This meant he did not arrive until January 1828, and all the rooms in college were filled.

Instead, he took lodgings above the tobacconist across the street. The owner had an arrangement with the college so that Christ's students rented rooms there, though this sometimes led to problems. The owner of the shop across from the tobacconist complained to the Master of the College that Christ's students kept knocking hats off passers-by in the street from their first floor windows with a horsewhip.

In November 1828, Darwin was able to move into a set of rooms in college, as he later recalled: "in old court, middle stair-case, on right-hand on going into court, up one flight, right-hand door & capital rooms they were." They were indeed capital, and newly discovered college record books reveal that Darwin's rooms were the most expensive range in the college at £15 per term. The record books also reveal that his college bills over three years amounted to about £700.

Left: A view of the Backs at Cambridge around 1840.

Right: Christ's College, Cambridge around the time Darwin was a student there. This engraving shows the view Darwin had when he lived in lodgings above the tobacconist.

At college he became close friends with his cousin William Darwin Fox, who may have introduced Darwin to the latest craze of collecting beetles. Darwin soon discovered several novel ways of procuring rare and unusual specimens. He had a special cabinet made to house his collection, and sent records of his captures to the well-known entomologist James Stephens, who published records for all of British entomology at that time. These were Darwin's first words in print, and so, even as an undergraduate, and in a very small way, Darwin had begun to contribute to scientific knowledge.

Darwin's interests in science became a lifelong devotion, though he continued to have a passionate liking for shooting. He avidly read the scientific travel accounts of Alexander von Humboldt, and dreamed of travelling to the Canary Islands on a scientific tour of his own. Another influential work for Darwin was by the astronomer John Herschel. His *Preliminary Discourse* (1831) became the authority on correct methods of scientific investigation. Darwin became the devoted pupil of John Stevens Henslow, professor of botany, from whom he learned a great deal about scientific method. The two became such good friends that college dons who had not met Darwin referred to him as "the man who walks with Henslow". No matter what he would become in later life, it was clear that Darwin would always maintain a strong interest and active participation in scientific research.

Left: The German naturalist and traveller Alexander von Humboldt. His book on South America exhilarated Darwin when he was a student in Cambridge, and filled him with a desire to emulate Humboldt.

Darwin's rooms at Christ's College, photographed in 1909.

TARGET PRACTICE

"When at Cambridge I used to practise throwing up my gun to my shoulder before a looking-glass to see that I threw it up straight. Another and better plan was to get a friend to wave about a lighted candle, and then to fire at it with a cap on the nipple, and if the aim was accurate the little puff of air would blow out the candle. The explosion of the cap caused a sharp crack, and I was told that the Tutor of the College remarked, 'What an extraordinary thing it is, Mr Darwin seems to spend hours in cracking a horse-whip in his room, for I often hear the crack when I pass under his windows.'" Charles Darwin, *Autobiography*, 1958, page 44.

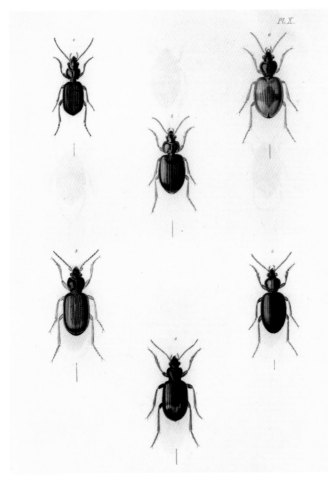

Pl. X.

CAPTURING BEETLES

Darwin later recalled one of his many beetling adventures: "one day, on tearing off some old bark, I saw two rare beetles and seized one in each hand; then I saw a third and new kind, which I could not bear to lose, so that I popped the one which I held in my right hand into my mouth. Alas it ejected some intensely acrid fluid, which burnt my tongue so that I was forced to spit the beetle out, which was lost, as well as the third one."

Hand-coloured engravings of British beetles from James Stephens'sß *British Entomology*. Darwin's first published words appeared in this work.

Darwin also studied other branches of natural science in his own time – as the university then offered little instruction in science – eventually learning the basics of a wide range of fields.

In 1831 he successfully completed his exam to gain the BA degree. He would later need to take special divinity training to become a clergyman. Darwin recalled in later life "Upon the whole, the three years I spent at Cambridge were the most joyful of my happy life."

Right: The astronomer John Frederick William Herschel.

Far right: John Stevens Henslow geologist, botanist and clergyman. He was Darwin's scientific mentor at Cambridge and recommended him for the *Beagle* voyage.

A page from the newly discovered Christ's College student bills for the "Quarter ending L. D. [Lady Day]" 1830. Darwin's £5 16s 4d for the grocer is more than that of his fellow students.

1830. Quarter ending L.D.

		Rickards	Dawson	Graham	Baker	Haworth	C Darwin	Davidson	
Dr.	Balance last Qr.	20.13.4				77.11.5			
	Bedmaker		1.1.–		1.1.–	1.1.–	1.1.–	1.1.–	
	Cash				12.		6.8		
	Coals		2.15.–	1.5.–	1.17.6	3.17.6	4.12.6	3.17.6	
	Cook	.14.7	4.10.5	2.4.5	6.	1.16.–	6.–.–	2.1.1	
	Laundress		2.4.8			2.4		2.	
	Porter	6.	9.2	9.5	1.3.10	7.4	1.4.10	13	
	Scullion	4.6	4.6	4.6	4.6	4.6	4.6	5.6	
	Semstress							1.	
	Shoeblack	7.	7	7	7	7.	7	7	
	Steward	4.5.7	7.16.10	6.1.9	5.8.11	13.3.10	15.18.2	8.10.11	
	Study Rent . . .	11.	5.	8.1.	3.15.	5.	4.	5	
	Tutor	2.10.	2.10.	2.10.	2.10.	2.10	2.10.	2.10.	
	College Account	29.12.0	26.18.7	20.3.1	34.7.9	103.2.7	36.4.8	26.7.0	
	Apothecary							1.15.6	
	Barber		.4.6				2.1.	.3.6	
	Bookseller								
	Circ. Library . .	7.6	.2.6						
	Brazier								
	Bricklayer		1		1		1	1	
	Carpenter					2.	1.6	1	
	Glazier		.3.6		.2.6	2.6	3.6	2.6	
	Grocer	4.10	4.3.7	2.6.10		2.3	3.13.4	2.5.10	
	Hatter				1.5.6				
	Linendraper . . .		2.2.3		6.3				
	Painter						3.9		
	Shoemaker	1.5.	15.6		13.6	1.3.6	1.18.	.10.	
	Smith								
	Tailor					6.7.6	3.15.	3.12.4	
	Upholsterer . . .					6.7.11			
	Private Tuition				14..	2.3.			
	Tradesmen's Bills	1.17.4	8.14.10	2.6.10	16.14.9	19.15.5	13.4.6	8.8.8	
	Total	31.9.4	35.13.5	22.9.11	51.2.6	127.18.0	49.9.2	34.15.8	
Cr.	Balance last Qr.								
	Scholarships . . .								
	Furniture								
	Total								
	Balance Dr. . . .								
	Balance Cr. . . .								
	* Received	30.–.–	35.13.5	22.9.11	51.2.6	127.18.–	49.10.–	34.15.8	
	* Paid								
	Balance Dr. . . .	1.9.4	✕✕✕	✕✕✕	✕✕✕	✕✕✕	✕✕✕	✕✕✕	
	Balance Cr. . . .		✕✕✕	✕✕✕	✕✕✕	✕✕✕	✕✕✕	✕✕✕	
	* Date		May 24,1830	May 10,1830	May 10,1830	May 21,1830	June 4,1830	May 14,1830	

A familiar letter from Darwin to his cousin William
Darwin Fox from 12 June 1828 – full of gossip and
details about collecting beetles.

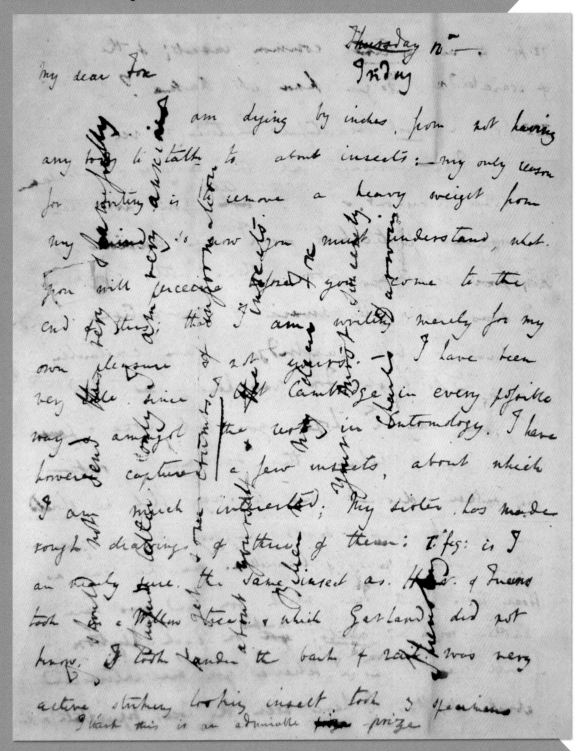

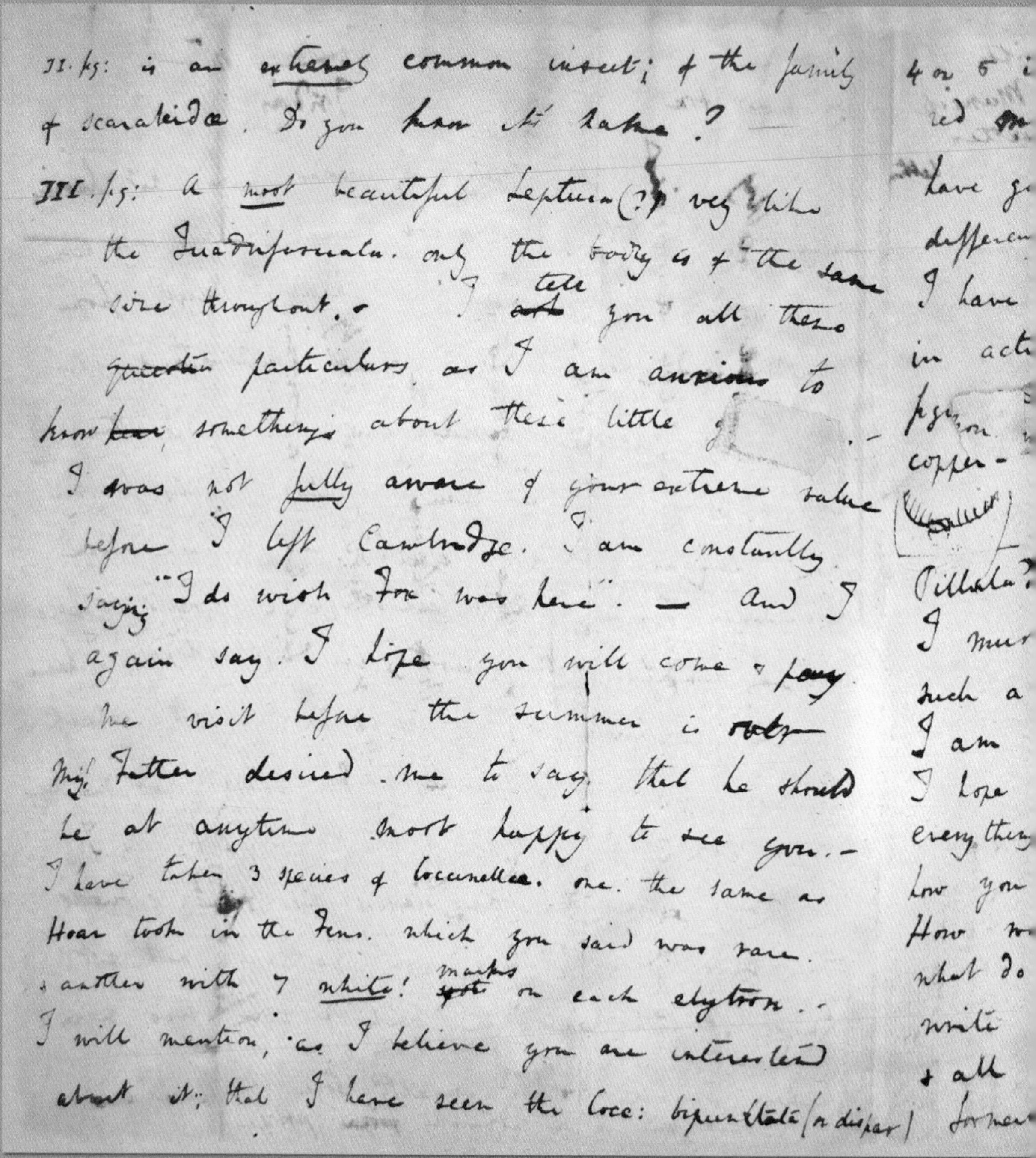

II. ys: is an extremely common insect; of the family
of scarabidæ. Do you know its name? —

III ys: A most beautiful Leptura (?) very like
the Quadrifasciata. only the body is of the same
size throughout. — I tell you all these
particulars, as I am anxious to
know something about these little g   . —
I was not fully aware of your extreme value
before I left Cambridge. I am constantly
saying "I do wish Fox was here". — And I
again say. I hope you will come & pay
me visit before the summer is over.
My Father desired me to say. that he should
be at anytime most happy to see you. —
I have taken 3 species of Coccinella. one. the same as
Hoar took in the Fens. which you said was rare.
& another with 7 white! marks on each elytron. .
I will mention, as I believe you are interested
about it; that I have seen the Coc: bipunctata (or dispar) for several

4 or 5 i
red on
have g
differen
I have
in acti
pigeon
copper-
Pittala?
I mus
such a
I am
I hope
everythi
how you
How m
what do
write
& all
for nea

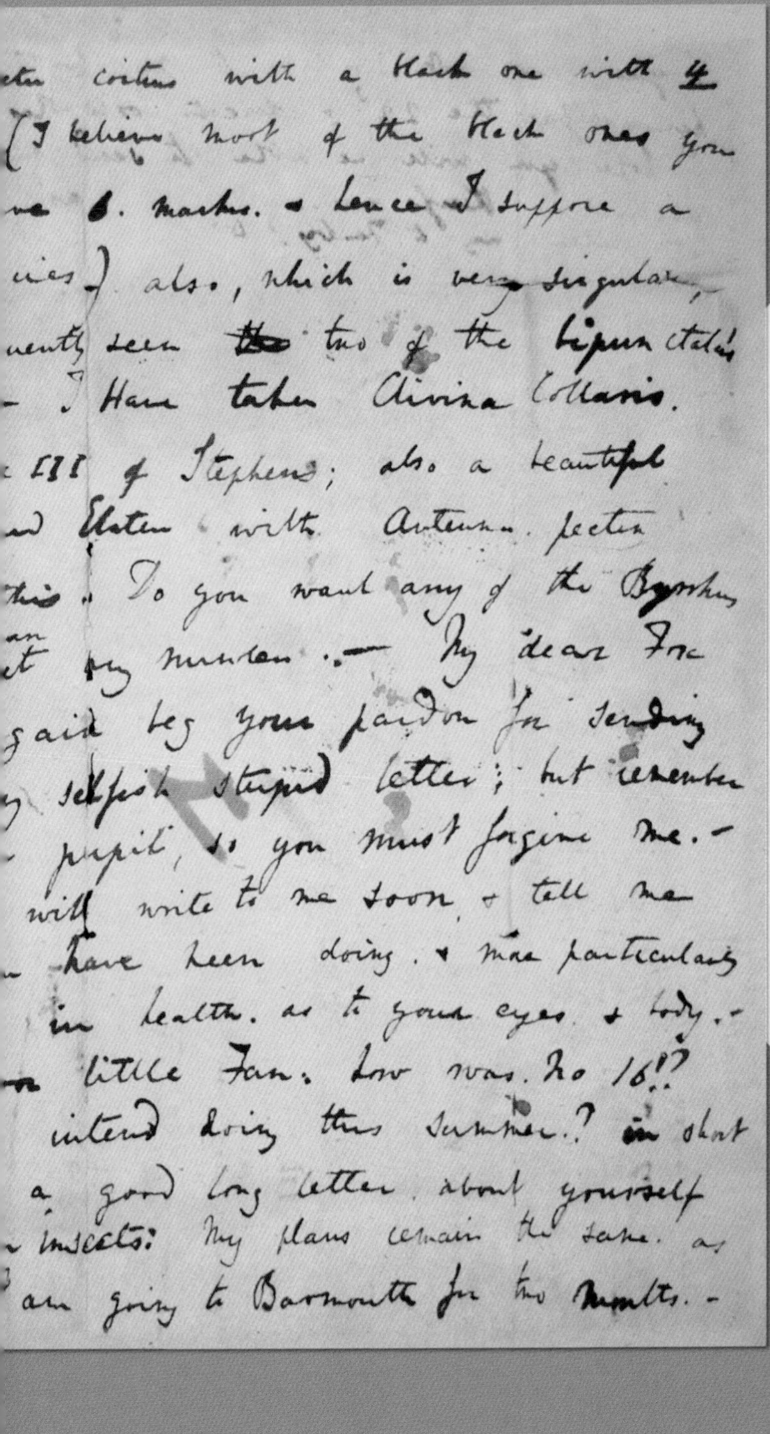

...ctus continues with a black one with 4
(I believe most of the black ones you
...ve 6. marks. — hence I suppose a
...cies) also, which is very singular,
...ently seen the two of the bipun ctati
— I have taken Clivina Collaris.

... 538 of Stephens; also a beautiful
...d Elster with Autumnal pecten
...this. Do you want any of the Byrrhus
...at by number. — My dear Fox
...gain beg your pardon for sending
...y selfish stupid letter; but remember
... pupil, so you must forgive me. —
...will write to me soon, & tell me
... have been doing. & more particularly
... in health. as to your eyes. & body. —
... little Fan. how was. No 16!?
... intend doing this summer.? in short
... a good long letter. about yourself
... insects: My plans remain the same. as
... am going to Barmouth for two months. —

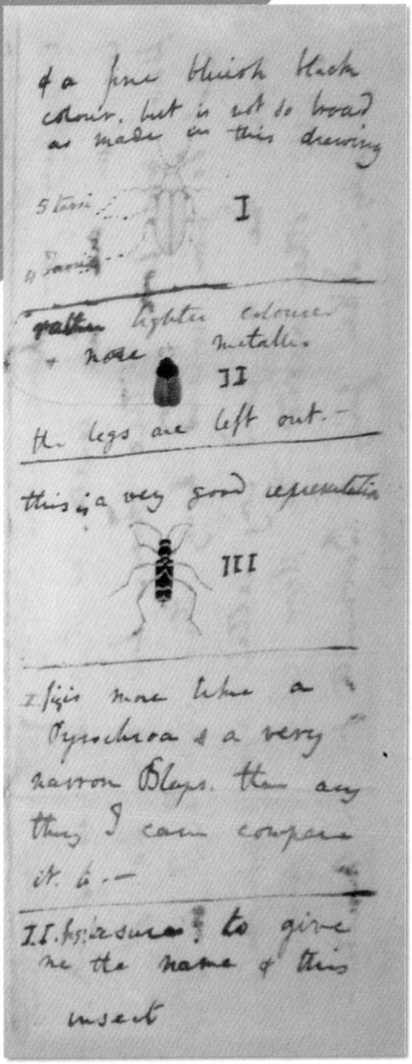

& a fine bluish black
colour. but is not so broad
as made in this drawing

5 tari....
6 tari....   I

rather lighter colour
+ nose   metallic
                II

the legs are left out. —

this is a very good representation
                III

I figis more like a
Pyrochroa & a very
narrow Blaps. than any
thing I can compare
it to. —

II. hp'æswre to give
me the name of this
insect

THE VOYAGE OF THE *BEAGLE*

In 1831, Darwin was fresh from university and budding with scientific talent. John Henslow, his Cambridge mentor could see this and encouraged Darwin to study geology, which Darwin took up with enthusiasm, contrary to his ealier declaration never to study the science.

Later, he accompanied Professor Adam Sedgwick on a geological tour of North Wales, where he was sent on a parallel track to Sedgwick's and each evening they compared notes. This helped Darwin learn the fundamentals of field geology first-hand. Darwin arrived home on 29 August to find a letter from Henslow awaiting him. It contained an offer that would change his life, and the world, forever.

A 26-year-old naval officer, Robert FitzRoy, had been given command of HMS *Beagle* for a second surveying voyage to South American waters. He was determined to take along a naturalist capable of studying the little-known lands the ship would visit.

He appealed to the Hydrographer of the Navy, Captain Francis Beaufort, to find such a person. FitzRoy wanted someone scientifically qualified, but it went almost without saying that he must also be a gentleman. A wide social gap existed between the captain and officers of a ship, and an independently wealthy gentleman would be a welcome companion on the long voyage. Beaufort contacted his friend George Peacock of Trinity College, Cambridge. Peacock first suggested the Reverend Leonard Jenyns, but he could not leave his parish. Peacock then consulted Henslow, who recommended his favourite pupil, Darwin. Darwin's father was against the plan, although he added "If you can find any man of common sense, who advises you to go, I will give my consent." Darwin wrote a letter to decline the offer before heading off to the house of his uncle Josiah Wedgwood II (Jos) for the start of the shooting season. Darwin's uncle and cousins thought such an opportunity as the *Beagle* was perfect for Darwin, as he was "a man of enlarged curiosity". Wedgwood sent for Darwin, who was out shooting, and together they drove to Shrewsbury to convince Darwin's father. Robert Darwin gave his consent, which meant not only that he was happy for Darwin to go, but that he would pay for the trip. Darwin wrote to Peacock and Beaufort to accept the offer and went to Cambridge to consult with Henslow before going on to London to meet FitzRoy.

FitzRoy proposed that he and Darwin should mess together aboard ship and that at any time during the voyage he could return home. As befitted the freedoms of a self-paying guest, Darwin was also free to keep the specimens he collected, in spite of the fact that specimens collected by naval officers were normally considered to be the property of the Government.

Opposite: Watercolour of the *Beagle* by Owen Stanley, 1841.

The *Beagle* was to survey the southern portion of South America and the Galápagos Islands, and to carry a chain of chronometric measurements back around the far side of the world. Darwin was not the only supernumerary on board: FitzRoy also engaged an artist and an instrument-maker to keep the ship's more than 20 chronometers in order. There were also a missionary and three Fuegians, whom FitzRoy had brought back from the previous voyage. In all, there would be 74 people on board a vessel 27m (90ft) long and 7.35m (24.5ft) wide amidships. Darwin would work and sleep in the 3 by 3.3m (10 by 11ft) poop cabin at the stern. It was dominated by a large chart table in the centre and lined with cupboards and bookcases.

After consulting with scientific experts and buying equipment ranging from a Bancks microscope and a set of aneroid barometers to a brace of pistols, in all costing £600, Darwin boarded the *Beagle* and opened a diary to record his experiences. After two false starts owing to bad weather, they set sail from Devenport on 27 December 1831. The world lay before them.

Below: Darwin's pocket sextant, possibly the one with which FitzRoy measured the height of a Baobob tree on St Jago.

The geologist Adam Sedgwick in 1850

A BURNING ZEAL

"During my last year at Cambridge I read with care and profound interest Humboldt's *Personal Narrative*. This work and Sir J. Herschel's *Introduction to the Study of Natural Philosophy* stirred up in me a burning zeal to add even the most humble contribution to the noble structure of Natural Science. … Henslow then persuaded me to begin the study of geology. Therefore on my return to Shropshire I examined sections and coloured a map of parts round Shrewsbury. Professor Sedgwick intended to visit N. Wales in the beginning of August to pursue his famous geological investigation amongst the older rocks, and Henslow asked him to allow me to accompany him." Charles Darwin, *Autobiography*, 1958, pages 67–9.

A YOUNG MAN OF PROMISING ABILITY

"Anxious that no opportunity of collecting useful information, during the voyage, should be lost; I proposed to the Hydrographer that some well educated and scientific person should be sought for who would willingly share such accommodations as I had to offer, in order to profit by the opportunity of visiting distant countries yet little known. ... Captain Beaufort approved of the suggestion, and wrote to Professor Peacock, of Cambridge, who consulted with a friend, Professor Henslow, and he named Mr Charles Darwin, grandson of Dr Darwin the poet, as a young man of promising ability, extremely fond of geology, and indeed all branches of natural history." Robert FitzRoy, *Narrative*, 1839, volume 2, page 18.

Robert FitzRoy, the *Beagle*'s commander. He and Darwin were affectionate friends during the voyage, apart from occasional outbursts because of FitzRoy's unpredictable temper.

Above: "Crossing the line" the traditional naval ceremony for seaman who first cross the equator as drawn by Augustus Earle. This is the only contemporary illustration from on board the *Beagle*.

"General chart shewing the principal tracks of H.M.S. *Beagle* 1831–6."
This map first appeared folded in a pocket inside the front cover of
FitzRoy's appendix to his *Narrative of the Surveying Voyages of His
Majesty's Ships Adventure and Beagle* (1839).

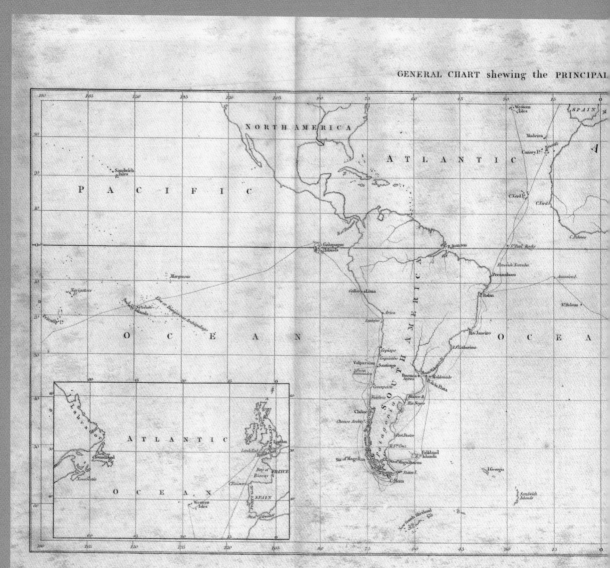

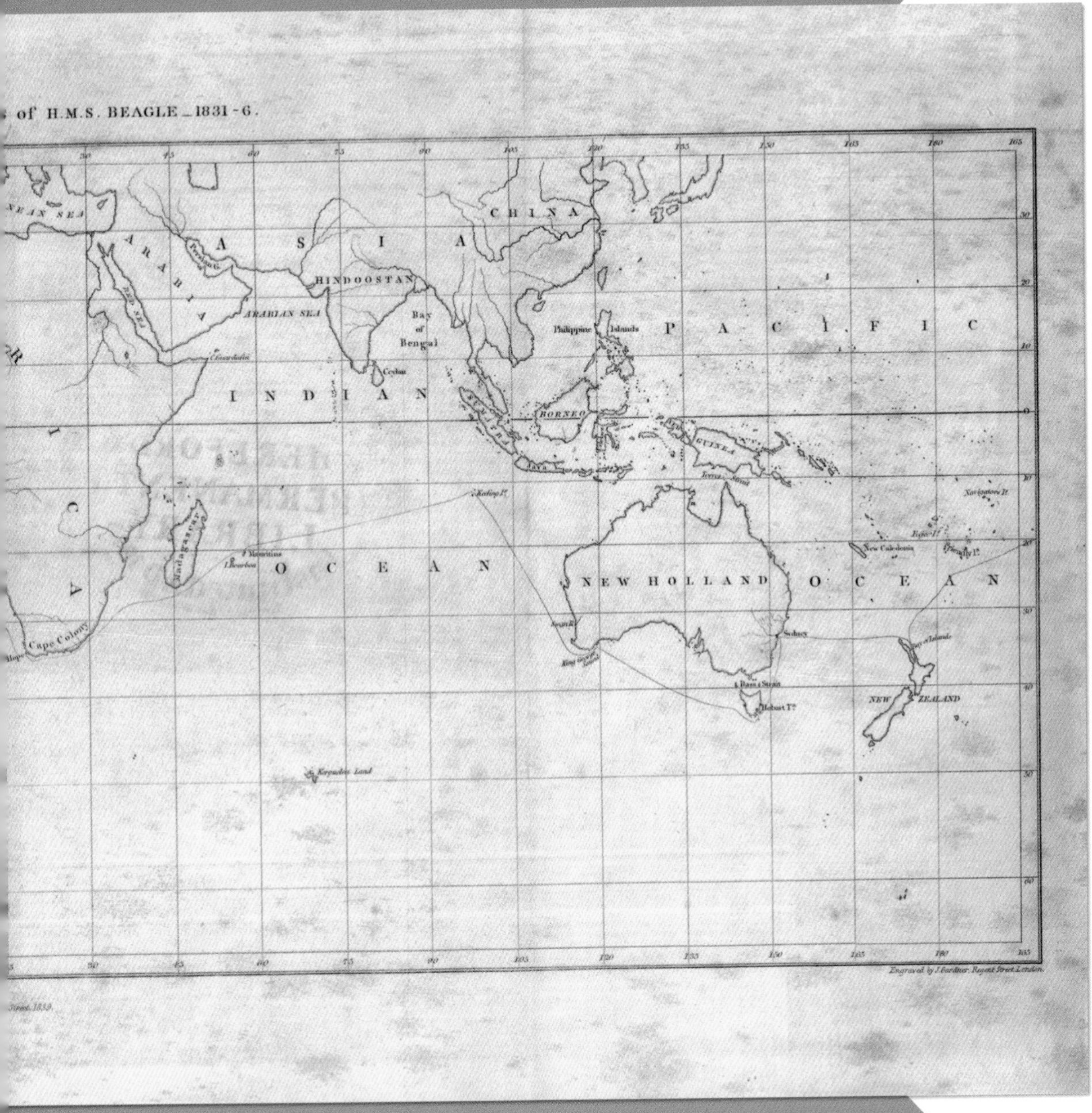

NEAN SEA

ARABIA

ASIA

CHINA

HINDOOSTAN

ARABIAN SEA

Bay
of
Bengal

Ceylon

Philippine Islands

PACIFIC

INDIAN

BORNEO

GUINEA

Keeling I?

OCEAN

NEW HOLLAND

O C E A N

New Caledonia

Cape Colony

NEW ZEALAND

Kerguelen Land

Engraved by J.Gardner, Regent Street London.

EASTERN SOUTH AMERICA

Darwin and Captain FitzRoy are often portrayed as antagonists during the voyage, but this is misleading. FitzRoy did become an evangelical Christian, but only after the voyage. It was FitzRoy who gave Darwin his copy of the first volume of Lyell's *Principles of Geology*, and Darwin was not inherently sceptical of the Bible.

In fact, he later recalled the officers of the *Beagle* laughing at him "for quoting the Bible as an unanswerable authority on some point of morality". Darwin's greatest misfortune in the early months at sea, and indeed throughout the voyage, was seasickness, which often left him incapacitated in his hammock.

The *Beagle*'s first stop was the Cape Verde islands 620km (385 miles) off the west coast of Africa. Here Darwin examined the geology of the island then called St Jago. He was thrilled to find the rocks "showed me clearly the wonderful superiority of Lyell's manner of treating geology, compared with that of any other author". The island had gradually changed over time. Firstly, Darwin could see that a stream of lava had once flowed over the sea bed – this baked the shells and corals there into a hard white rock – and, afterwards, these underwater rocks were pushed out of the sea. Yet the line of white rock also revealed subsidence around the volcanic craters on the island that had deposited yet more lava above. It was thrilling to be able to gaze so far into the past. Darwin sat in the shadow of an

Left: St Jago in the Cape Verde islands where the *Beagle* made her first landfall during the five-year round-the-world voyage.

outcrop of the white rock and thought of what he was able to prove about the history of the island with the evidence around him. At that moment he realized he could write a book about the geology of the lands visited by the *Beagle*.

The *Beagle* proceeded to Brazil where she was stationed for 19 days at Bahia (now known as Salvador, or Salvador da Bahia) where, for the first time, at the end of February 1832, Darwin experienced the breathtaking abundance of the tropics. The exotic fauna and flora and the astonishing variety of natural sights and sounds all around him was one of the most exhilarating experiences of his life. But he was soon hard at work. His pocket field notebook shows he was writing torrents of calculations, geological sections, measurements of angles, temperatures, barometer readings, compass bearings, diagrams and sketches.

A MEMORABLE HOUR

"The geology of St Jago is very striking yet simple: a stream of lava formerly flowed over the bed of the sea, formed of triturated recent shells and corals, which it has baked into a hard white rock. Since then the whole island has been upheaved. But the line of white rock revealed to me a new and important fact, namely that there had been afterwards subsidence round the craters, which had since been in action, and had poured forth lava. It then first dawned on me that I might perhaps write a book on the geology of the various countries visited, and this made me thrill with delight. That was a memorable hour to me, and how distinctly I can call to mind the low cliff of lava beneath which I rested, with the sun glaring hot, a few strange desert plants growing near, and with living corals in the tidal pools at my feet." Charles Darwin, *Autobiography*, 1958, page 81.

Opposite: "A view of Montevideo" by William Marlow. The largest city and chief port of Uruguay, which Darwin often called "Banda Oriental".

Above: Darwin's pocket telescope used on the *Beagle* voyage.

Right: Patagonian Indians in a "toldo" or skin tent. On the right is the tomb of a recently deceased child. Next to it are the stuffed skins of two sacrificed horses.

AN UNEXPLORED LAND

The South America visited by Darwin in the 1830s had only broken free from Spanish and Portuguese rule after a series of wars for independence between 1810 and 1825. With the opening up of South America to foreign trade, Britain was keen to chart the seas around it to enable profitable shipping. At the time, the interior of South America was less explored than that of Africa or Asia. The human slavery Darwin so abhorred was still common, and a chaotic mixture of bribery, extermination of indigenous peoples and profit through mining, cattle ranching and trade prevailed on the continent.

HMS *Beagle* in Straits of Magellan. Mount Sarmiento in the distance.

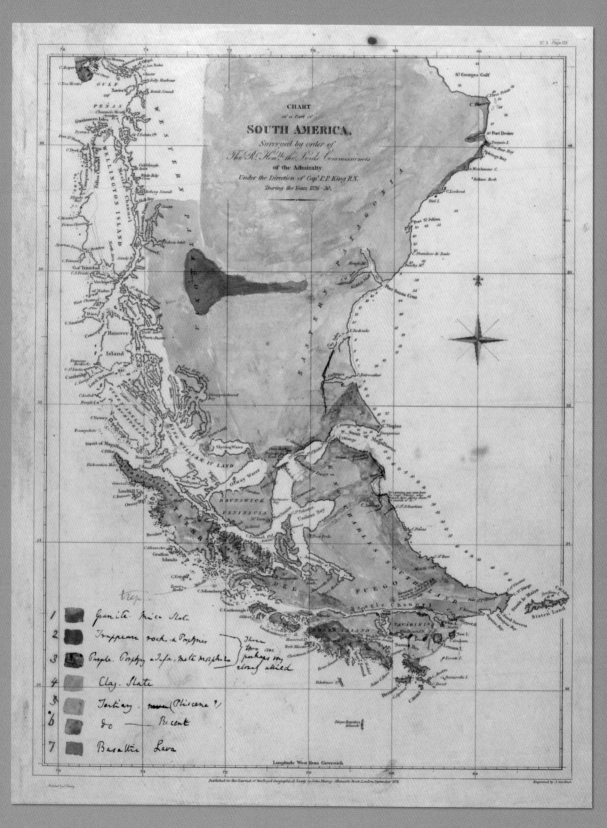

CHART
of a Part of
SOUTH AMERICA,
Surveyed by order of
The R.t Hon.ble the Lords Commissioners
of the Admiralty
Under the Direction of Cap.t P.P. King R.N.
During the Years 1826-30.

1 Granite. Mica Slate
2 Trappean rock, & Porphyry
3 Purple. Porphyry & Tufa. metamorphic
4 Clay. Slate
5 Tertiary. nerve (Pliocene?)
6 do ——— Recent
7 Basaltic Lava

These two are perhaps very closely allied

Published for the Journal of the Royal Geographical Society by John Murray. Albemarle Street. London, September 1832.

Engraved by J. Gardner.

Onboard ship he wrote up more formal notes on his specimens and collections. He created a system for cataloguing his finds so that the location and identity of each object would be recorded, making them identifiable once shipped back to England. He kept separate notes such as "Animals", "Reptiles in Spirits of Wine", "Fish in Spirits of Wine", "Insects in Spirits of Wine", and so forth. He kept large geological and zoological diaries, and on one exceptional day he collected 68 different species of beetle. This was a radically different world from the fens of Cambridgeshire.

Over the next two years, the *Beagle* proceeded to Rio de Janeiro, Montevideo, Bahía Blanca, Patagonia and the Falkland islands, all the while surveying the coasts and measuring the depth of the seas. Darwin spent about 1,162 nights onshore during the entire nearly five-year voyage and only 579 onboard ship.

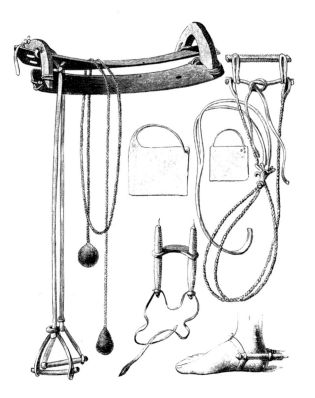

DON CARLOS "SO MUCH OF A GAUCHO"

Darwin repeated throughout his life that he was a poor linguist. He could not pronounce French or German and no doubt had a bad ear for accents. But this tends to obscure the fact that much of the voyage of the *Beagle* was experienced by Darwin in Spanish.

For weeks at a time he was deep in the interior of South America speaking with guides and staying with local people who spoke no English. Darwin had studied Spanish even before the voyage was offered to him and he learned as much as he could as the *Beagle* proceeded towards South America. He often recorded the name for birds or rocks in the local language and even read Spanish novels for light entertainment. The South Americans called him "Don Carlos", and he was introduced as un naturalista. What "a naturalist" was no one knew, but one local man was informed that it meant "a man that knows every thing".

Darwin spent a lot of time in the company of gauchos, South American cowboys in bright coloured ponchos with long knives in their belts. They taught Darwin a great deal about the local terrain and its wildlife. If Darwin knew that multiple female rheas, the South American ostrich, laid eggs in the nest of a single male, it was because the knowledgeable local gauchos told him so – in Spanish.

Darwin came to admire the free lifestyle of the gauchos, who were at home in the saddle across the vast plains of the Pampas. In the evening they would draw up their horses and declare, "here we will spend the night, under the broad open

Opposite, top left: A depiction of the traditional dress of a Gaucho by Juan Manuel Blanes.

Opposite, top right: Patagonian Ornaments and Riding Gear, from Ratzel's *The History of Mankind* (1904).

Opposite, bottom: Guachos using bolas to hunt rheas. Illustration by Frank Feller in *Chatterbox* (1894).

Right: The smaller species of rhea, or South American ostrich, which impressed Darwin as it was so similar to the larger species and their territories slightly overlapped.

JUAN MANUEL DE ROSAS (1793–1877)

On one long expedition between the Río Negro and Bahia Blanca, Darwin met the Argentine general Juan Manuel de Rosas, who was leading an unofficial war of extermination against the indigenous Indians. It was a dangerous time. The struggle was primarily over ownership of the land, though many of the confrontations took the form of brutal retaliation raids. Darwin was told that the Indians tortured all their prisoners and the Spaniards shot theirs. Rosas's troops had established a series of staging posts or *postas* along the main routes across the Pampas. Darwin and a few gaucho companions followed these perilous routes, always on the lookout for danger.

General Rosas gave Darwin permission to travel through the territory he controlled. Rosas later became the dictator of Argentina before being overthrown and retiring to England.

skies". They lived almost entirely on meat, and Darwin soon learned to adapt. Already a good rider and excellent marksman, he learned to drink their maté and smoke their small cigars. These he continued to enjoy for the rest of his life while being read to on the sofa by his wife after lunch. The gauchos also taught Darwin how to throw the bolas, although on his first attempt he entangled his own horse. "The Gauchos roared with laughter; they cried out that they had seen

every sort of animal caught, but had never before seen a man caught by himself."

We must try to imagine the young *Don Carlos, un naturalista*, galloping across the Pampas with pistols in his belt and geological hammer in his saddlebag, chatting in Spanish with his companions. But time and again he would pause to take notes in his pocket notebook, in English, with only a few Spanish words here and there. During one miserable rainy night in August 1833

THE VOYAGE OF THE *BEAGLE*

Throughout the voyage Darwin kept in contact with friends and family through letters. These sometimes took many months to reach their destination. In addition, Darwin periodically sent new portions of his personal diary home for his family to read. The diary was later used as the basis for Darwin's first book, *Journal of Researches into the Geology and Natural History of the Various Countries Visited by H.M.S. Beagle*. The book is now commonly called *The voyage of the Beagle*. Darwin also sent specimens home in boxes throughout the voyage to Henslow, who received and stored them in Cambridge until Darwin's return.

A caterpillar-hunting ground beetle collected by Darwin in Patagonia.

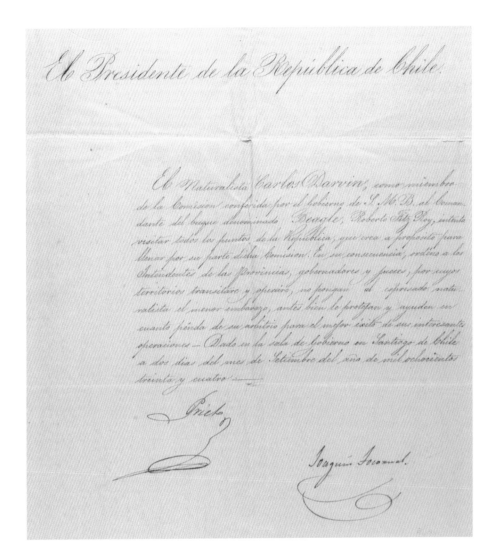

Right: Darwin's passport issued by the president of Chile. Darwin is called "El naturalista-Carlos Darwin".

Darwin jotted in his notebook: "Ship not arrived: very bad night, excessive rain so much of a gaucho do not care about it". For the rest of his life he would sprinkle his letters with Spanish sayings such as *adios*.

One of the great facts that impressed Darwin in South America, and which later led him to his theory of evolution, was the change of species as one moved southwards down the continent. The range of one species would finish and another, very similar species, would commence. An example is the rhea, or South American ostrich. Throughout the Pampas Darwin was familiar with the common rhea, but the gauchos told him of a smaller, rarer sort that they called the *Avestruz Petise*, which was seldom seen on the plains bordering the Río Negro. When camping near Port Desire, one of Darwin's companions shot a small rhea which they ate for dinner. Darwin at first assumed that the bird was an immature juvenile of the common sort. Only after dinner did he remember the rare *Petise*. He collected the remains: "head, neck, legs, wings, many of the larger feathers, and a large part of the skin". Darwin later learned that further south this kind of rhea was "tolerably abundant", in effect taking the place of the northern species. Why this should be he could not imagine, but it was a very curious fact and one that he continued to contemplate over the next few years.

FOSSIL DISCOVERIES

In the southern reaches of Patagonia, Darwin discovered the fossilized bones of giant extinct mammals. He spent many hours digging them out of river banks assisted by his servant Syms Covington. There were giant bones, teeth and mysterious armour plates, of which several later proved to be new to science.

Darwin could see immediately that some of the fossils resembled the unique present inhabitants of South America, such as armadillos and sloths. This was not unlike specimens found in Australia, where extinct fossil marsupial animals had been discovered. The concern for Darwin was why these creatures had become extinct, particularly since the Pampas were now so tranquil. One possibility was that a catastrophic flood had spread across the entire region, sweeping away these monstrous beasts.

One interesting source of evidence Darwin observed was the shells embedded in the earth with the extinct giants, which were virtually identical to those still living in the nearby sea. He also found fossilized whale bones with fossil barnacles attached to them. Through careful examination of the evidence, Darwin discovered that the whale bones had been exposed for some time in the sea – long enough to become encrusted with barnacles – before they were covered with sediment. This showed that the fossil mammals were not wiped out by a catastrophic flood, but that their carcasses had been carried out to an estuary and, in a quiet environment, gradually covered with silt. The same process was still happening. A great flood would have wiped out the shells as well as the mammals and not allowed time for the growth of barnacles. This, however, did not explain what had extinguished the giant mammals.

The bones had lain on the eastern coast,

which had gradually become submerged by the sea. They were now visible on dry land because there must have been a subsequent gradual elevation of the land. Some of this uplifted land was later cut through by rivers, which allowed Darwin to find them.

Fortunately, Darwin had recently received the second volume of Lyell's book on geology, which discussed the extinction of species. Lyell proposed that species seemed to disappear piecemeal in the fossil record, probably because of local natural causes, and that new species were created to replace them. How this happened was left quite vague, but readers were left to assume that it was by an act of creation. Darwin now had the opportunity to face real examples relevant to the latest theoretical questions.

The fossils themselves were all described by Richard Owen after the voyage in *Zoology of the Voyage of H.M.S. Beagle* (1838–43), edited and supervised by Darwin. Were it not for the community of scientists debating and publishing on these subjects, Darwin could never have made sense of his fossil finds – they would simply have been mysterious old bones. But in the context of the science of his day, it was possible to understand that the remains were from very ancient animals (by human standards), now vanished from the earth never to return and strangely similar to the present animals of South America as opposed to any others in the world.

Opposite, top: The *Beagle* laid ashore for repairs on the Santa Cruz river, as depicted by Conrad Martens.

Opposite, bottom: Patagonians at Gregory Bay by Conrad Martens. Published in FitzRoy's *Narrative* in 1839.

THE *MEGATHERIUM*

The *Megatherium* (Latin for "huge beast") was studied and named by Georges Cuvier in 1796. A complete skeleton had been shipped to Madrid from South America, and Georges Cuvier showed that, despite it being the size of a small elephant, the animal belonged to the same family as the living sloths peculiar to South America. It was clear that such a large animal could not have gone unobserved for centuries, and so it became one of the earliest creatures known to be extinct. In Darwin's day the Megatherium was one of only three known large fossilized species found in South America.

A reconstructed skeleton of a *Megatherium*. Darwin found many bones of this extinct giant ground sloth.

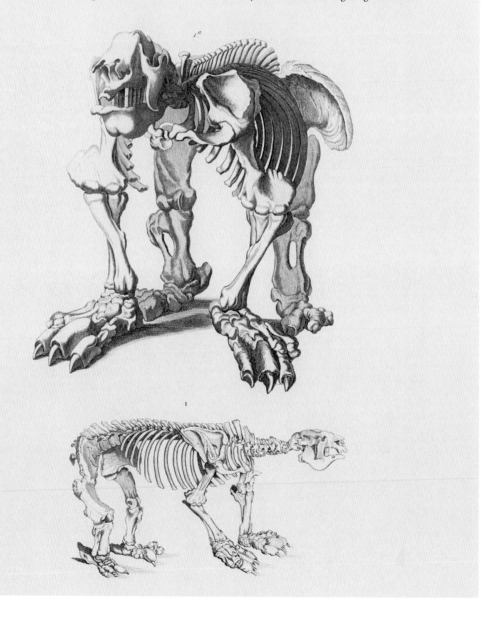

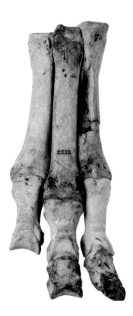

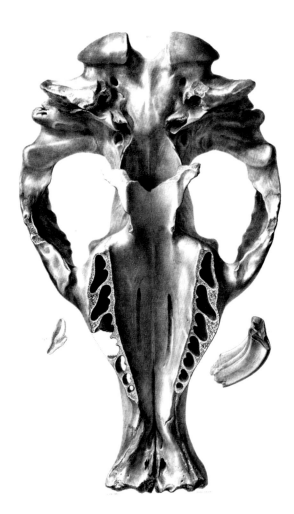

Above: Bones of the right forefoot of Darwin's specimen of *Macrauchenia patachonica*, now in the Natural History Museum in London.

Above right: Darwin's discovery of the teeth of an extinct horse in ancient rocks proved that horses had once existed in the New World.

Right: Base of the skull of *Toxodon platensis* (the genus name means Archer's bow teeth) found by Darwin about 193km (120 miles) from Monte Video. This engraving was reproduced life-size over 60cm (two ft) long in *Zoology of the Beagle* (1838).

The skull of a *Toxodon platensis* collected by Darwin near Montevideo.

BURIED REMAINS

In a preface to his *Zoology of the Voyage of H.M.S. Beagle*, Darwin wrote: "we may feel certain, that at a period not very remote, a great bay occupied the area both of the Pampas and of the lower parts of Banda Oriental. Into this bay the rivers … must formerly have carried down (as happens at the present day) the carcasses of the animals, inhabiting the surrounding countries; and their skeletons would thus become entombed in the estuary mud which was then tranquilly accumulating. Nothing less than a long succession of such accidents can account for the vast number of remains now found buried."

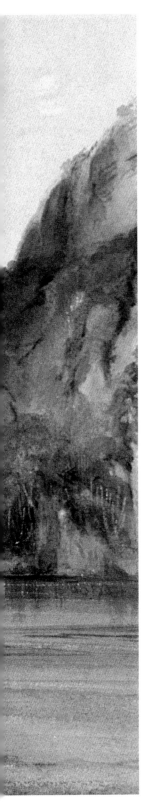

TIERRA DEL FUEGO AND THE SHOCK OF THE SAVAGE

On his previous voyage, Captain FitzRoy had taken four Indians from Tierra del Fuego, the archipelago at the southern tip of South America, back to England to be educated and Christianized with the intention of returning them to their homeland.

The three young men were known as Boat Memory, York Minster and Jemmy Button, and the young girl, Fuegia Basket. Although Boat Memory died from smallpox in England, the remaining three were aboard with Darwin on the *Beagle* for a time, along with a young English missionary, Richard Matthews, who planned to stay with them and found a mission. Darwin barely mentioned the Fuegians in his diary of the early voyage except once to mention that Fuegia Basket "daily increases in every direction except height."

When the *Beagle* encountered almost-naked Fuegians in December 1832, Darwin was deeply shocked. "It was without exception the most curious & interesting spectacle I ever beheld. – I would not have believed how entire the difference between savage & civilized man is. – It is greater than between a wild &

Left: Watercolour of HMS *Beagle* in the Murray Narrows, Beagle Channel, Tierra del Fuego, painted by Conrad Martens and purchased by Darwin in Australia in 1836.

domesticated animal." The Fuegians, despite the cold temperatures, stood almost naked gathered on the beach to see the foreigners. To Darwin's ears their language sounded like throat clearing. They had few possessions, just spears and bows and arrows. "The skin is dirty copper colour … the only garment was a large guanaco skin, with the hair outside. – This was merely thrown over their shoulders, one arm & leg being bare … Their food chiefly consists in limpets & muscels [sic], together with seals & a few birds." On another occasion Darwin observed "a woman, who was suckling a recently-born child … whilst the sleet fell and thawed on her naked bosom, and on the skin of her naked child." The difference between these wild people and the polite and well-groomed Jemmy Button, with his polished shoes, was astonishing.

A mission was built with three wooden wigwams, and plenty of stores and gardens were planted. The local Fuegians begged for every object they saw and stole whatever they could. Even so, the *Beagle* left Matthews and the three Anglicized Fuegians as it went to carry out further surveying work, and returned after two weeks. Matthews almost leaped into the sea to return to the *Beagle*. At the moment the ship had appeared, Fuegians were plucking out his beard one hair at a time using mussel shells as pincers. The mission was a failure, but the Anglicized Fuegians had no wish to return to England.

A year later, in early 1834, the *Beagle* returned to the site. Jemmy had been robbed and abandoned by York Minster, who took Fuegia Basket away with him. The *Beagle*'s crew were dismayed to see Jemmy had reverted to "savagery", and was almost naked with long tattered hair.

The experience of seeing these wild hunter gatherers who slept curled up on the ground like animals had a profound effect on Darwin. It finally convinced him of the undeniable animality of human beings. Civilization, though apparently so vastly different from the wildness before him, was just a veneer that could vanish with altered circumstances, as when the well-dressed Jemmy was returned to the wilds of Tierra del Fuego. Nothing could have brought more powerfully before Darwin's mind the actual range of human variation. His own species ranged at the present day from

Left: A man of the Tekeenica tribe by Conrad Martens. Darwin was shocked to see in what animal-like conditions other humans could live and survive.

Opposite, bottom: Fuegian wigwam at Hope Harbour in the Magdalen Channel, by Philip Parker King.

apparently wild savages to the most cultivated English high society. But underneath it all, Darwin could see that humans were still animals who ate, slept, killed and infected each other with diseases.

A NEAR MISS

The *Beagle* braved the stormy and dangerous waters rounding Cape Horn. On 13 January 1833, during a gale with force 11 winds, a series of monstrous waves struck the *Beagle*. The first checked her speed, the second turned her from the wind and lengthways to a third mountainous wave. When it struck, the ship was pushed over right on its side. Half the deck was covered by water and if a fourth wave had followed, the *Beagle*, and Charles Darwin, would have been no more. FitzRoy recalled: "For a moment, our position was critical; but, like a cask, she rolled back again, though with some feet of water over the whole deck."

Sorely Tried, HMS *Beagle* off Cape Horn, 13 January 1833 at 1:45 p.m., by John Chancellor.

A GRIEVOUS CHANGE

After a year the *Beagle* revisited Jemmy Button. "It was quite painful to behold him; thin, pale, & without a remnant of clothes, excepting a bit of blanket round his waist: his hair, hanging over his shoulders; & so ashamed of himself, he turned his back to the ship as the canoe approached. When he left us he was very fat, & so particular about his clothes that he was always afraid of even dirtying his shoes; scarcely ever without gloves & his hair neatly cut. – I never saw so complete & grievous a change." *Charles Darwin*, Beagle *Diary*, edited by R D Keynes, 1988, page 226.

FitzRoy published these phrenologically informed portraits of native Fuegians they encountered. The ones in European dress were some of those taken to England by FitzRoy during the first *Beagle* voyage. This image appears in FitzRoy's *Narrative*, 1839.

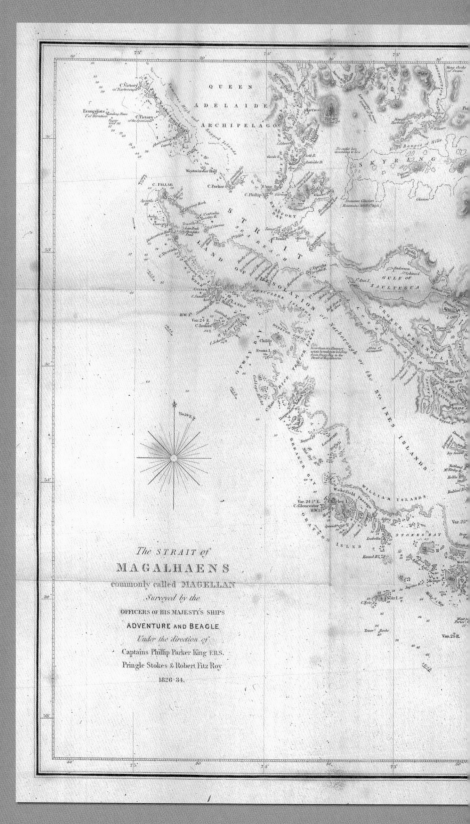

A survey map of the Straits of Magellan by the officers of the *Adventure* and *Beagle* from Philip Parker King's Volume of *Narrative* (1839). The small numbers along the coast indicate soundings in fathoms taken by the crew.

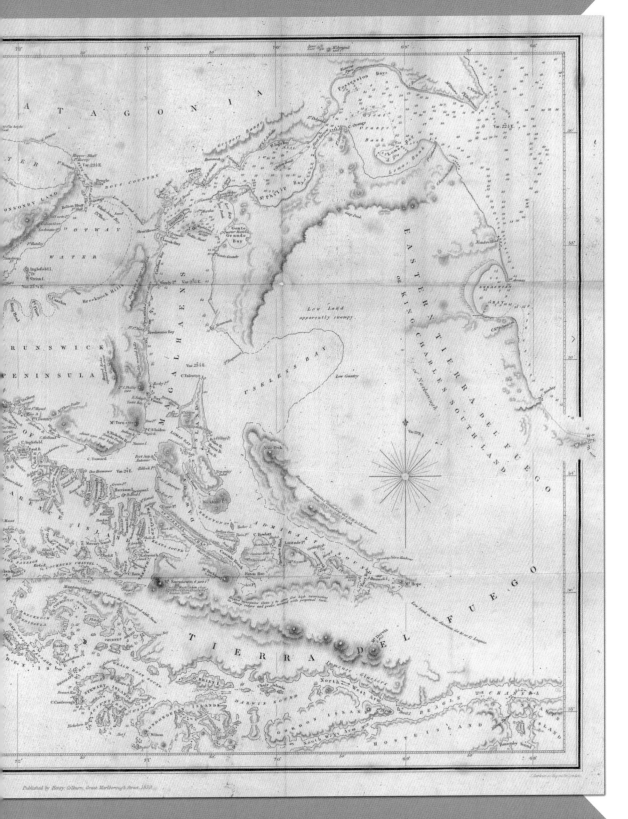

Published by Henry Colburn, Great Marlborough Street, 1839.

WESTERN SOUTH AMERICA

From the middle of 1834, the *Beagle* surveyed the western coast of South America, making frequent calls in Chile and Peru. It was here that Darwin experienced first-hand some of the geological forces that change the surface of the earth, exactly the kinds of actual causes favoured by Lyell.

In early 1835, he witnessed two volcanoes erupting powerfully at night. The sea was illuminated with a long bright shadow and in the midst of the glare of the eruption large dark stones could be seen shooting into the air before falling to the earth.

As well as destructive tidal waves, on another occasion Darwin experienced a tremendous earthquake which destroyed scores of towns and villages in Chile. Darwin recorded: "An earthquake like this at once destroys the oldest associations; the world, the very emblem of all that is solid, moves beneath our feet like a crust over a fluid; one second of time conveys to the mind a strange idea of insecurity, which hours of reflection would never create." The *Beagle* visited the port of Concepción in Chile. The earthquake and accompanying tsunami had devastated the city, its port of Talcuhano and 70 nearby villages. The great cathedral lay in ruins. Darwin noted that the walls of buildings perpendicular to the supposed direction of the shock were thrown down whereas many of those in line with it remained standing.

In the days and weeks after the earthquake, FitzRoy and Darwin pieced together what had happened. The earthquake had affected an area of 1,126km by 643km (700 by 400 miles), and repeated aftershocks came roughly from the east, sometimes leaving long north-to-south cracks in the ground. Darwin recorded in his diary "The Earthquake & Volcano are parts of one of the greatest phenomena to which this world is subject." The shoreline, as could be seen from previously submerged rocks, now exposed, and high-water lines of mussels, was elevated 2.4m (8ft) above its previous level. Darwin began to search for inland beds of marine shells as evidence of previous earthquakes. He found what he expected to find and continued to find them 70m

Opposite: An engraved sketch by Conrad Martens of San Carlos de Chilóe, from FitzRoy's *Narrative* 1839.

Right: Old Wooden Church at Castro, drawn by Philip Parker King, from FitzRoy's *Narrative* 1839.

7 DECEMBER 1834

"In the evening we reached the island of S. Pedro, where we found the *Beagle* at anchor. In doubling the point, two of the officers landed to take a round of angles with the theodolite. A fox, of a kind said to be peculiar to the island, and very rare in it, and which is an undescribed species, was sitting on the rocks. He was so intently absorbed in watching their manœuvres, that I was able, by quietly walking up behind, to knock him on the head with my geological hammer. This fox, more curious or more scientific, but less wise, than the generality of his brethren, is now mounted in the museum of the Zoological Society." Charles Darwin, *Journal of Researches,* 1839, page 341.

Left: The Chilotan fox (*Canis fulvipes*), which Darwin felled with his geological hammer on 7 December 1834.

(230ft) above the sea. The local people did not believe the shells were marine because they were located in the mountains. "It is amusing to find the same subject discussed here as formerly amongst the learned of Europe concerning the origin of these shells, whether they were really shells or were thus 'born by Nature'."

Applying his training in Lyell's principles of geology, Darwin found that the western coast of South America was gradually being uplifted by natural causes still in operation, and that so comparatively small an event as the earthquake he had experienced, if reiterated over a long enough period of time, was sufficient to explain the mighty mountain chain of the Andes.

The certainty of the continued elevation and subsidence of parts of the Earth's crust

brought Darwin surprisingly close to modern plate tectonics. What eluded him was any sense of the crust also drifting horizontally. The principle of elevation and subsidence over vast areas also enabled Darwin to come up with one of his first great theories – coral reefs and atolls are formed.

This understanding, however, did not always lead to successful theories. Darwin later applied the principle of vertical subsidence and uplift to explain the remains of beaches along mountainsides in Scotland ("Glen Roy"). He was convinced that the principles he had found so often in South America and the Pacific would also explain these features. He was later very disappointed with his theory when it was shown that glaciers had blocked up the valleys permitting the formation of freshwater lakes.

Opposite: A map of the island of Chiloé and parts of the adjacent coasts from HMS *Beagle*, as published in Fitz Roy's volume of *Narrative* (1839).

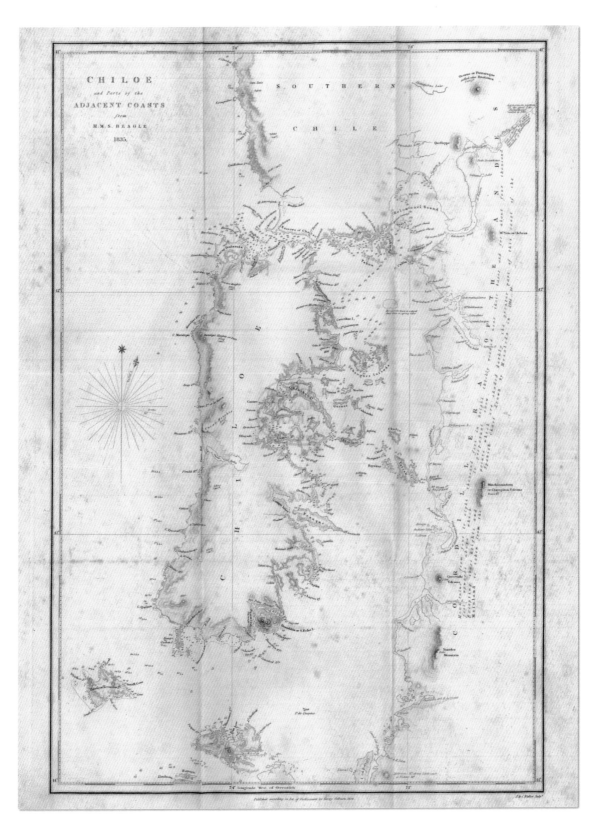

CHILOE
and Parts of the
ADJACENT COASTS
from
H.M.S. BEAGLE
1835.

DARWIN'S THEORY OF CORAL ATOLLS

"No other work of mine was begun in so deductive a spirit as this; for the whole theory was thought out on the west coast of S. America before I had seen a true coral reef. … I had during the two previous years been incessantly attending to the effects on the shores of S. America of the intermittent elevation of the land, together with denudation and the deposition of sediment. This necessarily led me to reflect much on the effects of subsidence, and it was easy to replace in imagination the continued deposition of sediment by the upward growth of coral. To do this was to form my theory of the formation of barrier-reefs and atolls." Charles Darwin, *Autobiography*, 1958, page 98.

Valdivia: "the town is situated on the low banks of the stream, and is so completely buried in a wood of apple-trees that the streets are merely paths in an orchard."

Right: "View in the Cordillera" by John Miers (1826). Darwin often consulted the work of men like Miers, a mining engineer who travelled in South America in the 1820s and 1830s, to see what this earlier traveller had observed.

Below: Lima, Peru, which Darwin visited in July 1835. Photographed circa 1860.

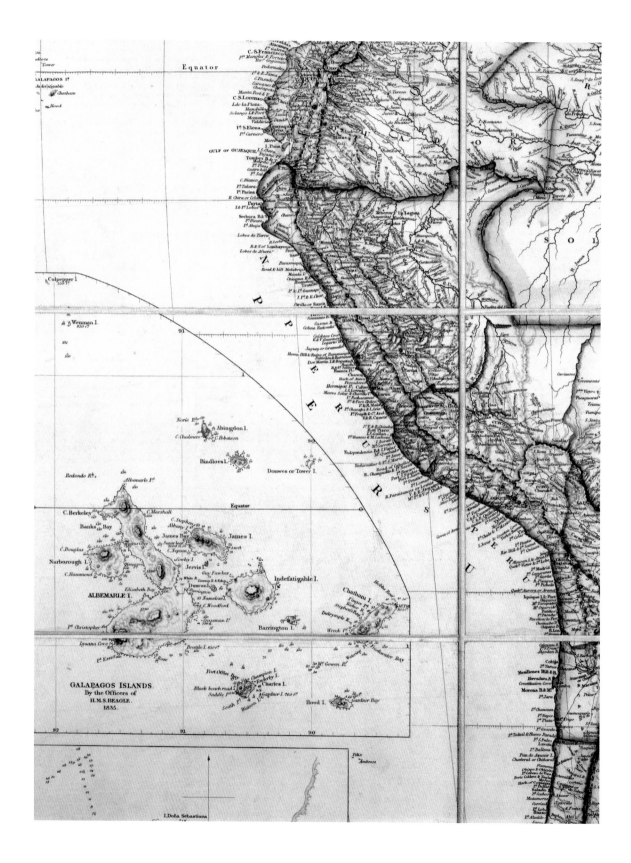

GALAPAGOS ISLANDS.
By the Officers of
H.M.S.BEAGLE.
1835.

THE GALÁPAGOS: THE TRUE STORY

No part of the *Beagle* voyage is today more famous or more shrouded in legend than the time in the Galápagos Islands. The *Beagle* was there for five weeks, from 15 September to 20 October 1835, and crew members made a series of charts that were still in use in the 1940s.

Darwin spent about 19 days ashore, on Chatham, Charles, Albemarle and James Islands. Historians now know that Darwin did not discover evolution while in the islands. In fact, the author's research has shown that the legend that Darwin discovered evolution while actually in the Galápagos, and because of his observations of the beaks of the finches, did not arise until the mid-twentieth century.

When the *Beagle* left the west coast of South America for the last time, Darwin was particularly interested in studying the geology of the Galápagos. But when the *Beagle* arrived, Darwin was not impressed with the largely barren and rocky islands. He soon learned, however, that the islands were of comparatively recent volcanic origin. The *Beagle*'s soundings revealed that the ocean was extremely deep

Opposite: Map of South America drawn by officers of HMS *Beagle*. Published in King's volume of *Narrative* (1839).

Right: HMS *Beagle* in the Galápagos, 17 October 1835 at 2:15 p.m., by John Chancellor. The *Beagle* is sending boats to fetch Darwin and company from the Galápagos for the last time.

THE DISTRIBUTION OF ORGANIC BEINGS

"I never dreamed that islands, about fifty or sixty miles apart, and most of them in sight of each other, formed of precisely the same rocks, placed under a quite similar climate, rising to a nearly equal height, would have been differently tenanted; but we shall soon see that this is the case. It is the fate of most voyagers, no sooner to discover what is most interesting in any locality, than they are hurried from it; but I ought, perhaps, to be thankful that I obtained sufficient materials to establish this most remarkable fact in the distribution of organic beings." *Charles Darwin, Journal of Researches*, second edition, 1845, page 394.

The Governor's well or "dripstone" at the base of a hill on Charles Island, Galápagos.

around and between the islands, which seemed to indicate that they were very tall volcanic mountains and not a visible extension of the continent of South America.

This made the islands' inhabitants all the more curious for Darwin. He could see that the islands had erupted as molten lava from the bottom of the sea, particularly when he found fossilized seashells in some deposits. This meant that when the islands first appeared they were devoid of life. While on the islands Darwin still believed in a version of Charles Lyell's views that species were created in a particular centre and could radiate outwards from there. He wondered which centre of creation the species on the Galápagos had migrated from.

The creatures of the Galápagos were famously unafraid of man. A gun was hardly necessary to collect specimens. Darwin even approached a hawk, which did not take flight, until he stood close enough to push it off its branch with the barrel of his gun.

The birds, in particular, were obviously like those in South America – though Darwin had not visited the part of the continent at the same latitude as the islands. Darwin could not help noticing that the mockingbirds on three different

Left: A specimen of *Nosomimus trifasciatus* or Charles mockingbird, collected in the Galápagos islands. It was the mockingbirds, rather than the finches, which attracted Darwin's particular attention in the Galápagos.

Right: Galápagos mockingbird (*Mimus melanotis*) collected on James Island. Darwin noted at the time that there were distinct mockingbirds on different islands – rather like the tortoises as he had been informed.

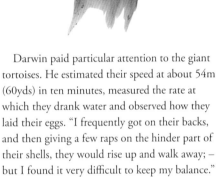

islands were distinct, and he noted this fact on the specimens he collected. It did not occur to him at the time that many of the land birds would differ on different islands, since many of the islands were in sight of one another. Darwin did not even know, for example, that all of the kinds of what, since 1935, have been called "Darwin's finches", were finches at all. Until he had his specimens checked by an expert ornithologist with a worldwide collection, he could not determine if his specimens were distinct species or only local varieties.

MYSTERY OF MYSTERIES

"The natural history of these islands is eminently curious, and well deserves attention. Most of the organic productions are aboriginal creations, found nowhere else … yet all show a marked relationship with those of America. … The archipelago is a little world within itself … Seeing every height crowned with its crater, and the boundaries of most of the lava-streams still distinct, we are led to believe that within a period, geologically recent, the unbroken ocean was here spread out. Hence, both in space and time, we seem to be brought somewhat near to that great fact — that mystery of mysteries — the first appearance of new beings on this earth." *Charles Darwin, Journal of Researches,* second edition, 1845, page 377.

Darwin testing the speed of a Galápagos tortoise. Darwin's dress and facial hair here are typical of the 1890s, when this was drawn for a popular biography by Charles Frederick Holder.

Darwin paid particular attention to the giant tortoises. He estimated their speed at about 54m (60yds) in ten minutes, measured the rate at which they drank water and observed how they laid their eggs. "I frequently got on their backs, and then giving a few raps on the hinder part of their shells, they would rise up and walk away; – but I found it very difficult to keep my balance."

There were two species of iguana of a type found only on the Galápagos. One species was terrestrial. He observed them making their burrows in the sand. "I watched one for a long time, till half its body was buried; I then walked up and pulled it by the tail; at this it was greatly astonished, and soon shuffled up to see what was the matter; and then stared me in the face, as much as to say, 'What made you pull my tail?'"

As part of his studies he dissected some of the marine iguanas, unique in the world with their semi-aquatic lifestyle, and found that they were herbivores.

Even though he was not stirred to evolutionary speculation on the Galápagos, they were later to influence his thinking profoundly, and would provide one of the three main inspirations for Darwin's theory of evolution.

ACROSS THE PACIFIC AND AROUND THE WORLD

From the Galápagos the *Beagle* set sail ever westwards travelling 5,150km (3,200 miles), past the Low or Dangerous Archipelago, where Darwin "saw several of those most curious rings of coral land, just rising above the water's edge, which have been called Lagoon Islands", to Tahiti.

Darwin and FitzRoy found the natives pleasant and intelligent, and Darwin went on an expedition into the mountains with Tahitian guides who showed him how they lived off the land. Both were impressed by the activities of the missionaries, and FitzRoy had an official meeting with the Queen who was later invited aboard the *Beagle*, where she was entertained with singing and rockets.

The *Beagle* next sailed for New Zealand – where Darwin was much less impressed with the natives, but again was impressed with the missionaries' activities – before sailing on to Sydney, Australia. During their stay Darwin did not see much of the famous wildlife. He recorded in his diary "I had been lying on a sunny bank & was reflecting on the strange character of the Animals of this country as compared to the rest of the World. An unbeliever in everything beyond his own reason, might exclaim 'Surely two distinct Creators must have been [at] work; their object however has been the same & certainly the end in each case is complete'. – Whilst thus thinking, I

observed the conical pitfall of a Lion-Ant: … Without a doubt this predacious Larva belongs to the same genus, but to a different species from the European one. – Now what would the Disbeliever say to this? Would any two workmen ever hit on so beautiful, so simple & yet so artificial a contrivance?"

Aborigines showed Darwin how they threw boomerangs and spears with throwing sticks. Darwin's genteel English sensibilities were, however, shocked by the rough Australian society with its ex-convict servants and wealthy upstart entrepreneurs. He penned in his diary as the *Beagle* sailed on "Farewell Australia, you are a rising infant & doubtless some day will reign a great princess in the South; but you are too great & ambitious for affection, yet not great enough for respect; I leave your shores without sorrow or regret."

They called in at the Keeling Islands in the Indian Ocean, where Darwin had the opportunity to carefully examine a coral atoll. On they sailed to Mauritius and from there to South Africa in May 1836. As intended, they

Opposite, top: The missionary Henry Nott's old Chapel in Tahiti. From *Narrative* volume 2.

Opposite, bottom left: One of the fortified hill tops in New Zealand from the 1890 edition of *Journal of Researches*.

Opposite, bottom right: A Tahitian man as depicted in the 1890 edition of *Journal of Researches*.

MASTER OF MANY FIELDS

Darwin's reading, collecting, dissecting, describing and cataloguing skills were like the equivalent of taking modern advanced degrees in half a dozen scientific disciplines. In his work he ranged from the examination of microscopic polyps to the surveying and reconstructing of the geological history of entire continents, gaining a masterful degree of competence across many fields. This, and the fact that he was able to travel around the globe, meant that the distribution of life on Earth was laid out before him in a way others could only imagine from reading.

Darwin also studied ocean life. This image shows a group of corals. Illustration drawn by C Berjeau circa 1870.

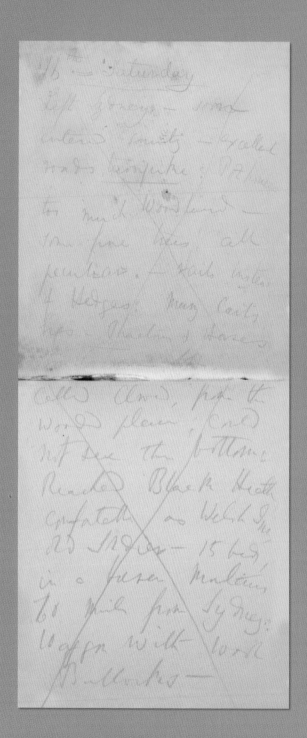

Extracts from one of Darwin's field notebooks used during the voyage of the *Beagle*. This portion records Darwin's January 1836 excursion to Bathurst, about 200km (125 miles) west of Sydney, Australia.

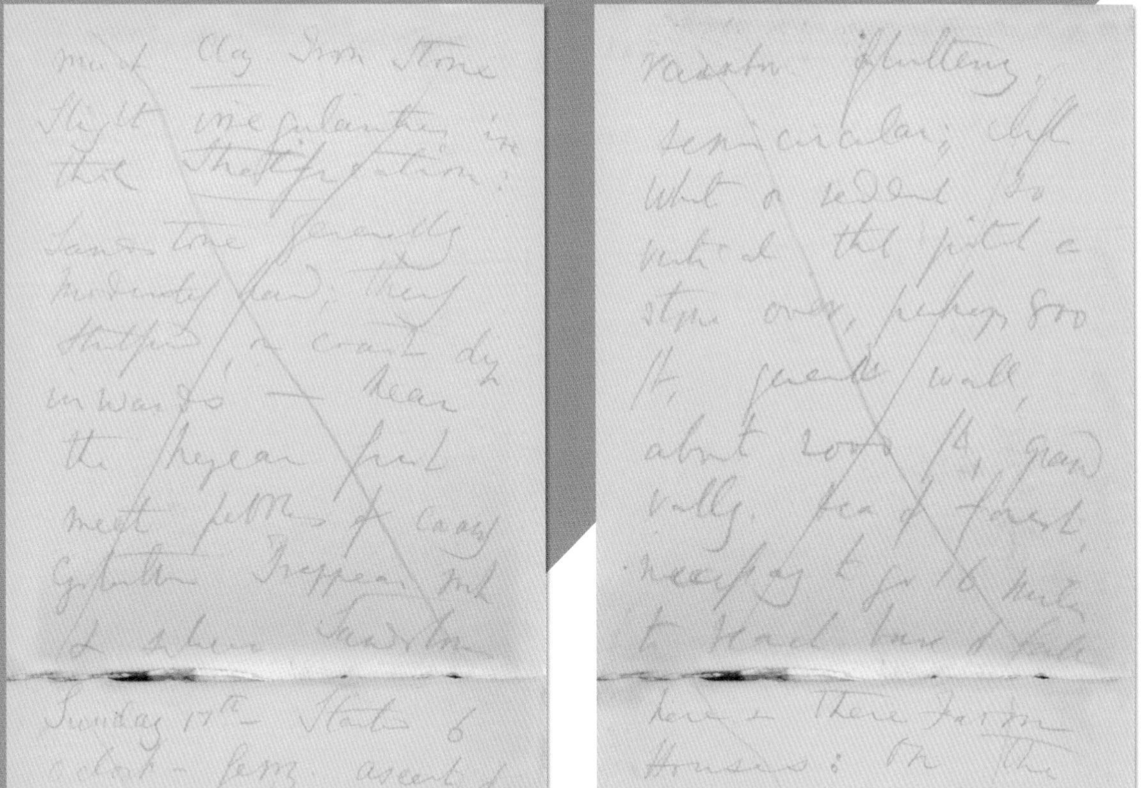

much clay Iron Stone
slight irregularities in
the stratification:
Sandstone generally
moderately hard; they
stratified or crack dry
in wards — near
the ?? feel
meet pebbles of coarse
Gippsland Trappean rock
& where Sandstone

Sunday 17th — Start 6
o'clock — began ascent of
Blue Mountains, great
fine Woods. —
Hen plain, uneven, many
valleys; gradually we
great deception, when
the elevation — considerable
of nearly 3000 ft —
Bare woods; piles
pale & peculiar green

vacation fluttering;
semicircular; cliff
What a sudden so
? the ? a
steep over, perhaps 800
ft, grand wall,
about 2000 ft, grand
valley, sea of forest,
necessary to go 16 miles
to reach bare & halt
here & there a few
Houses: on the
heyean, ? some
escarpment of Blue
mountains; contemplate
edge of great plateau;
cultivated land?
Part of Back Mrs,
beautiful precipices
in ?; ??
Speak English — None

Below right: Maoris as depicted in *Narrative* volume 2. FitzRoy believed the Maoris and the Tahitians were of the same race "descended from the same original stock".

Opposite, top: The *Beagle* at Port Louis, Mauritius, from the 1890 edition of *Journal of Researches*. Darwin was impressed by the grandeur of the island.

Below: Darwin's pocket clinometer used during the voyage to measure the angle or slope of a line of sight above or below horizontal or the height of an object. Made by Cary in London it cost Darwin 25 shillings.

stayed for a month to take chronometrical readings, and Darwin did some exploring. The scarcity of the vegetation reminded him of his giant South American fossils. If elephants and rhinoceroses could subsist on the African savannahs, then his South American giants would not have vanished for lack of food. Darwin was also thrilled to meet one of the most eminent English scientists of the day, the astronomer Sir John Herschel, who was then living in South Africa, to make astronomical observations.

It was time for the *Beagle* to head home. They sailed to St Helena, where the remains of Napoleon still rested, thence to Ascension Island where they received some mail. "I received a letter whilst at Ascension, in which my sisters told me that Sedgwick had called on my father and said that I should take a place among the leading scientific men." Darwin was stunned to learn that his scientific labours had become known at home through his Cambridge mentor John Henslow, who had read some of his letters at a scientific meeting and had them printed. The *Beagle* continued sailing west again to Brazil, where Darwin enjoyed his last sight of tropical scenery before the *Beagle* departed for England, arriving home on 2 October 1836.

HOMEWARD BOUND.

THE STABILITY OF SPECIES

Between South Africa and England Darwin reflected on his findings and wrote the following in his bird notes: "When I recollect, the fact that [from] the form of the body, shape of scales & general size, the Spaniards can at once pronounce from which Island any Tortoise may have been brought. When I see these Islands in sight of each other, & possessed of but a scanty stock of animals, tenanted by these birds, but slightly differing in structure & filling the same place in Nature, I must suspect they are only varieties. ... If there is the slightest foundation for these remarks the zoology of Archipelagoes — will be well worth examining; for such facts [would] undermine the stability of Species."

A depiction of the *Beagle* homeward bound first published in the 1890 edition of *Journal of Researches*.

TO MARRY OR NOT TO MARRY

As soon as the *Beagle* returned, Darwin rushed home to Shrewsbury to see his family where there was a joyous reunion "after an absence of 5 years & 2 days". It might have seemed that his adventure had come to an end; however, this was not the case.

Packed away in boxes, wrapped in paper or pickled in bottles of alcohol were thousands of specimens collected all around the world. His mentor John Henslow had stored them in Cambridge, and he told Darwin it would take twice as long to describe them in writing as it did to collect them. Darwin at first thought this was absurd, but it proved to be prophetic advice.

After several trips between Shrewsbury, Cambridge and London, Darwin took lodgings in December 1836 at a house on Fitzwilliam Street, Cambridge. He frequented the courts of his old college, Christ's, but found it wasn't the same now that the bright young faces that emerged from the archways were unfamiliar to him. The fellows at least were keen to hear Darwin's stories of adventure and he often dined with them in college.

In March 1837, Darwin moved to London to be closer to the scientific societies that were discussing his findings, such as the Geological Society. He and his family never made a conscious decision that he would not become a clergyman, instead his scientific life gradually filled all of his time. The important thing was that he had not become an idle sporting man, as his father had once expressed a concern he might, but instead was a respected figure in the British metropolitan scientific scene.

Darwin's collections and papers proved to the scientific elite that young Mr Darwin was a naturalist of the highest calibre. He was also well-liked not only because of his good humour and cheery disposition but also because he was both very intelligent and able, yet entirely lacking in arrogance or off-putting ambition. Living on a generous allowance from his father he was, in every sense, a gentleman of science.

In addition to his other theorizing at this time, Darwin began to think seriously about religion. He read anthropologists' accounts of peoples from around the world and over the course of history. Darwin reasoned that, since almost all of them believed in what were by then considered to be false religions, and since these religions that had spread far and wide without any evidence to substantiate them, Christianity might be just one among all the rest. A traditional argument for the existence of God – that a belief in God is innate in all men – also lost its power over him. Anthropology showed that there were peoples in the world with no such belief.

Darwin also began to think about marriage. In his usual way he jotted notes to help his thinking. As advantages to not marrying he counted the "Freedom to go where one liked –

choice of Society & little of it. – Conversation of clever men at clubs – Not forced to visit relatives, & to bend in every trifle". His pro-marriage points included: "Children –(if it Please God) – Constant companion, (& friend in old age)". In the end he decided marriage was the best option and discussed the idea with his father, who advised him to conceal his religious doubts as they would only give pain to a wife.

Darwin settled on his cousin Emma Wedgwood, daughter of the Wedgwood pottery family, and, to his joy, Emma accepted. They were married in January 1839 and shortly moved into their first house on Gower Street in London. The garish colours of the house led them to dub it Macaw Cottage.

The marriage was a happy one, but, unfortunately, Darwin's health soon began

to deteriorate and Emma would care for him throughout his life. Their first son, William, was born in December 1839, and, not surprisingly, this latest specimen was carefully observed and noted by an affectionate Darwin.

Above: Somerset House, where Darwin read several important papers to the Geological Society.

Opposite, bottom: The library of the Athenæum Club. Darwin often used the library before his marriage.

Right: The Athenæum Club, Pall Mall. Darwin was elected a member in 1838.

THE ONSET OF DOUBT

"I had gradually come, by this time, to see that the Old Testament from its manifestly false history of the world, with the Tower of Babel, the rainbow as a sign, etc., etc., and from its attributing to God the feelings of a revengeful tyrant, was no more to be trusted than the sacred books of the Hindoos, or the beliefs of any barbarian. … the

clearest evidence would be requisite to make any sane man believe in the miracles by which Christianity is supported, – that the more we know of the fixed laws of nature the more incredible do miracles become, – that the men at that time were ignorant and credulous to a degree almost incomprehensible by us, – that the Gospels cannot be proved to have been written simultaneously with the events, – that they differ in many important details, far too important as it seemed to me to be admitted as the usual inaccuracies of eye-witnesses … I gradually came to disbelieve in Christianity as a divine revelation. … Thus disbelief crept over me at a very slow rate, but was at last complete. The rate was so slow that I felt no distress, and have never since doubted even for a single second that my conclusion was correct." *Charles Darwin,* Autobiography, 1958, page 85.

A copy of the New Testament in German signed "C Darwin H.M.S. *Beagle*", owned The Charles Darwin Trust.

Darwin's notes on the prospect of marriage and what it would mean in terms of his work. In this thoughtful analysis he carefully lists the advantages and disadvantages of marriage. These notes were written before Darwin became engaged to Emma and were probably written in July 1838.

Not Marry

o children, (no second life) no one to care for
one in old age. — What is the use of working
without sympathy from near & dear friends —
who are near & dear friends to the old. except
relatives — Freedom to go where one liked —
choice of Society & little of it. — Conversation
of clever men at clubs — Not forced to
visit relatives, & to bend in every trifle —
to have the expense & anxiety of children —
perhaps quarrelling — Loss of time. — cannot
read in the Evenings — fatness & idleness —
Anxiety & responsibility — less money for books &c —
if many children forced to gain one's bread. —
(But then it is very bad for one's health to work too much)
Perhaps my wife won't like London; then
the sentence is banishment & degradation
into indolent, idle fool —

JOURNAL OF RESEARCHES

INTO THE

GEOLOGY

AND

NATURAL HISTORY

OF THE

VARIOUS COUNTRIES
VISITED BY H. M. S. BEAGLE,
UNDER THE COMMAND OF CAPTAIN FITZROY, R.N.
FROM 1832 TO 1836.

BY

CHARLES DARWIN, Esq., M.A. F.R.S.

SECRETARY TO THE GEOLOGICAL SOCIETY.

LONDON:
HENRY COLBURN, GREAT MARLBOROUGH STREET.

1839.

JOURNAL OF RESEARCHES

Late in the voyage of the *Beagle*, FitzRoy asked to read part of Darwin's journal (or diary as it is now called to avoid confusion with the title of his first book).

FitzRoy declared that it was worth publishing and in the end it was proposed that Darwin write a third volume to the official narrative of the first and second voyages. It was entitled *Narrative of the Surveying Voyages of His Majesty's Ships Adventure and Beagle*, and was edited by FitzRoy. While settled in Cambridge in 1836-37 Darwin began adapting his ship-board diary into a book. Much of the text remained unchanged, though Darwin added condensed descriptions of some of his more interesting scientific findings. He completed the work in November 1837. FitzRoy, on the other hand, was nowhere close to completion so the set was delayed. Only in the middle of 1839 was it published and offered for sale.

Darwin's volume carried the subsidiary title *Journal and Remarks* and was so well written and interesting that it was soon offered for sale as a stand-alone volume with the title *Journal of Researches into the Geology and Natural History... .* The book made Darwin a considerable scientific celebrity, and the prestigious Quarterly Review described it as "One of the most interesting narratives of voyaging that it has fallen to our lot to take up, and one which must always occupy a distinguished place in the history of scientific navigation." The second edition, of 1845, transposed "Geology" and "Natural History"

in the title, and *The Voyage of the Beagle* was only given as a title to a 1905 edition.

While working on this book, Darwin first began to think systematically about the origins of species, but few of his dawning speculations made it into the almost completed text. For example, when discussing some of the small birds of western South America and its nearby

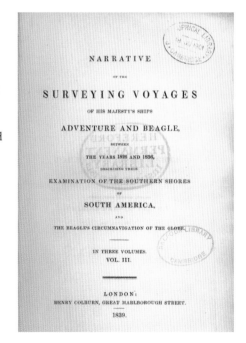

NARRATIVE

OF THE

SURVEYING VOYAGES

OF HIS MAJESTY'S SHIPS

ADVENTURE AND BEAGLE,

BETWEEN

THE YEARS 1826 AND 1836,

DESCRIBING THEIR

EXAMINATION OF THE SOUTHERN SHORES

OF

SOUTH AMERICA,

AND

THE BEAGLE'S CIRCUMNAVIGATION OF THE GLOBE.

IN THREE VOLUMES.
VOL. III.

LONDON:
HENRY COLBURN, GREAT MARLBOROUGH STREET.
1839.

Above: A Galápagos marine iguana from the second edition of *Journal of Researches* (1845).

Left: A Galápagos tortoise from the 1890 illustrated edition of *Journal of Researches*.

islands, he remarked "When finding, as in this case, any animal which seems to play so insignificant a part in the great scheme of nature, one is apt to wonder why a distinct species should have been created." Or when discussing the Andes as a great barrier separating species on the east and west coasts: "Unless we suppose the same species to have been created in two different countries, we ought not to expect any closer similarity between the organic beings on the opposite sides of the Andes than on shores separated by a broad strait of the sea."

By 1845, however, Darwin's views had matured and some of his new thinking about species appeared in the second edition prepared in that year. For example, when discussing the relationships between some of the extinct fossil giants and "the living sloths, ant-eaters, and armadillos, now so eminently characteristic of South American zoology", he concluded, "This wonderful relationship in the same continent between the dead and the living, will, I do not doubt, hereafter throw more light on the appearance of organic beings on our earth, and their disappearance from it, than any other class of facts." He described from the Galápagos, "a

AN IMPORTANT GEOLOGICAL PAPER

Within three months of stepping off the *Beagle* Darwin was standing before scientific societies in London, such as the Geological, Zoological and Entomological societies, describing his findings. On 7 March 1838, Darwin read one of his most important geological papers to the Geological Society. He argued that the progressive long-term changes to the geology of South America were due to incremental, non-catastrophic causes. In 1840 the revised paper was published as "On the connexion of certain volcanic phenomena in South America; and on the formation of mountain chains and volcanos, as the effect of the same powers by which continents are elevated".

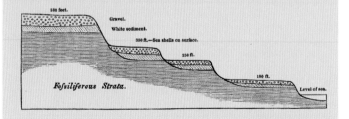

A section through the "step-like plains" of Patagonia showing the results of Darwin's geological investigations from *Journal of Researches* (1839)

A STATEMENT ON THE ORIGIN OF SPECIES

In his paper "On certain areas of elevation and subsidence in the Pacific and Indian oceans, as deduced from the study of coral formations", read before the Geological Society on 31 May 1837, Darwin made his first public (and published) statement of his interest in the origin of species: "That some degree of light might thus be thrown on the question, whether certain groups of living beings peculiar to small spots are the remnants of a former large population, or a new one springing into existence."

Below: Galápagos finches from the second edition of *Journal of Researches* (1845). "The most curious fact is the perfect gradation in the size of the beaks in the different species of Geospiza…"

Bottom: Engraving of the undersides of four flatworms collected and drawn by Darwin during the *Beagle* voyage, published in an article in 1844.

most singular group of finches, related to each other in the structure of their beaks, short tails, form of body, and plumage: there are thirteen species…" He added that "The most curious fact is the perfect gradation in the size of the beaks" as there were "no less than six species with insensibly graduated beaks". Darwin declared: "Seeing this gradation and diversity of structure in one small, intimately related group of birds, one might really fancy that from an original paucity of birds in this archipelago, one species had been taken and modified for different ends."

In this important scientific paper read before the Geological Society of London in 1837, Darwin made his first published reference to his interest in the origin of species.

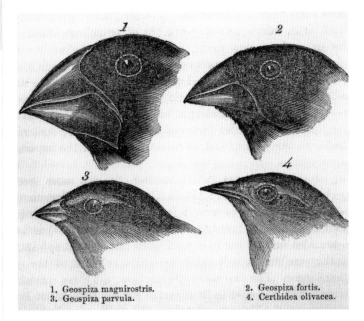

1. Geospiza magnirostris.
2. Geospiza fortis.
3. Geospiza parvula.
4. Certhidea olivacea.

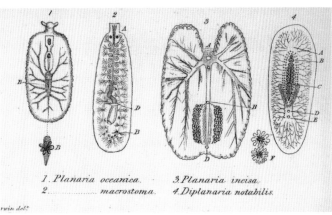

1. *Planaria oceanica.*
2. *—— macrostoma.*
3. *Planaria incisa.*
4. *Diplanaria notabilis.*

Pages from Darwin's pocket diary covering January 1839–September 1846 during which time he published the results of the *Beagle* voyage and his "Glen Roy" paper. The entry for 29 January 1939 reads "married & returned to London".

1839
Jan 11th Went to Shrewsbury
15 Went to Maer
18 London
25 Shrewsbury
28 Maer
29 Married & returned to London
Feb 7th Recommenced Coral Paper —
End of February & first week in March
Earthquake paper. then a Maer
week on Species & the Coral Paper
April 24 Went to Maer
May 13 to Shrewsbury
— 20 to London.
Aug 23 to Maer & thence on the
26 to Birmingham for the meeting of the Brit. Assoc.

First week January correcting Glen Roy Paper. — Did nothing during the rest of Month. —
Feb 5th Began German.
End of March & nearly all April Coral Paper. —
Maer visit. some reading connected with species, but did very little on account of being unwell.
May 20 to July 30. Map for Coral is. — ornithological part of Beagle Voyage — Lost some time unwell. read little for Species: to end of 23 of Aug. Coral. — May. & Hort at M.

1841 ... March 2 Annie born
paper on Boulders & Ice of S. America. — finished April 4th. — Was idle & unwell. sorted papers on Species Theory. May 28th went to Maer & Shrewsbury; returned to London July 23. July 26th commenced Coral work. after more than 13 month interval of Coral
Jan 3 1842 Sent M.S. to Printers.
March 7th went to Shrewsbury for 10 days:
May 6 corrected last proof of Coral Volume. — I commenced the work 3 years & 7 months ago. Out of this period (besides work on Beagle voyage) about 20 months has been spent on it, besides it, I have only completed the Bird Part of Zoology: appendix to Journal Paper on Boulders & corrected paper on Glen Roy &

March 2. 1841 Annie Born. —

Reading on Species & but also lost by illness. —

1842

May 18th Went to Maer. June 15th to Shrewsbury; & on 18th to Capel-Curig, Bangor Carnarvon to Capel-Curig, altogether ten days, examining glacier action. During my stay at Maer & Shrewsbury, wrote pencil sketch of my species theory. — July 18th returned to London. Wrote paper on Glaciers — Employed about Down. — Emma came to Down September 14th. & I followed on 17th. — Mary Eleanor Sept 23rd born. — Ob. October 16th. — October 14th began on Volcanic islands — to shorten & rearrange Coral glossary in S. —

1842. Sept. 14. Came to Down.

1846. Oct 1st Finished last proof of my Geolog. Obser. on S. America. This volume, including Paper in Geolog. Journal on the Falkland Islands took me 18 & ½ months: the M.S. however, was not so perfect as in case of Volcanic Islands. So that my Geology has taken me 4 & ½ years: now it is 10 years since my return to England. How much time lost by illness!
Oct 1. Paper on new Balanus ... 10 days in London during 2 visits ... November, December Corona & Megatrema

1846
Feb 21st to Shrewsbury. March 3rd Home
July 31st to do Aug 9th. Home
September 9th with Emma to Brit. Assoc: at Southampton, on the 12th to Portsmouth a coast of Isle of Wight on 13th to Winchester & S. Cross. on 14th Netley Abbey & Southampton Common. 17th Home
Sept. 22nd With Em. & Susan to Knole Park

ZOOLOGY OF THE *BEAGLE*

In early 1837, Darwin considered asking for government aid to publish the zoological findings of the voyage as a book.

By May his connections and reputation gained him the support of some prominent men of science, and on 16 August Darwin called on the Chancellor of the Exchequer, who gave the good news that the Lords Commissioners of the Treasury had granted £1,000 towards the costs of the book. Much of this paid for the expensive but sumptuous engravings used to illustrate the work, which is known as *The Zoology of the Voyage of H.M.S. Beagle* (1838–43).

Darwin convinced five distinguished experts to scientifically classify, name and describe his zoological specimens from the voyage. There would be five parts, including one written by Richard Owen, professor of anatomy, called *Fossil mammalia*. In this work Darwin's fossil giants, such as the *Macrauchenia* from South America, were named and described, and Darwin added a geographical introduction. Owen found that the specimens were not only very large, but all herbivores, and are curiously similar in type to those still found in South America. The principal bones were beautifully engraved including a life-size plate of the skull of the *Toxodon platensis*.

George Robert Waterhouse, curator of the Zoological Society, wrote *Mammalia*, which described the living mammals collected by Darwin and "their habits, ranges, and places of habitation" and, again, a geographical

Hand-coloured engraving of Felis Yagouaroundi from Brazil. This specimen was given to Darwin by a priest who had hunted it with dogs.

Left: Hand-coloured engraving of two species of South American mice captured by Darwin.

Opposite: Hand-coloured engraving of cactus-feeding finches (*Cactornis scandens*), by John Gould.

introduction was written by Darwin. It was illustrated with 35 hand-coloured plates of bats, foxes, mice and a dolphin named after FitzRoy. Waterhouse found that the molars of the South American rodents were different from Eurasian rodents, and so put them in a separate category. Darwin noted that the Tucutuco, a burrowing rodent from Brazil, was often blind, and added "it appears strange that any animal should possess an organ constantly subject to injury".

The ornithologist John Gould wrote *Birds*, which was illustrated with 50 magnificent hand-coloured plates, sketched by Gould himself, and which his wife engraved on stone. All but six were natural size. Gould classified Darwin's Galápagos finches not as varieties but as 13 distinct species. Before the volume was finished Gould left for an expedition to Australia, and Darwin completed the work with the help of George Robert Gray, the ornithological assistant in the Zoological department of the British Museum.

The cleric and naturalist Leonard Jenyns wrote *Fish*. Because Darwin's fish were preserved in alcohol, their colours were faded. Darwin, knowing this when he collected them, recorded their colours using the standard colour names in *Werner's Nomenclature of Colours* (1821), by Patrick Syme; however, the 29 plates illustrating the text were not coloured. This was to be Darwin's only work devoted to fishes.

The dental surgeon and naturalist Thomas Bell wrote the final part, *Reptiles*, which also included amphibians. Unfortunately, Bell

CATALOGUING SPECIES

Over the years of the *Beagle* voyage, Darwin had collected thousands of specimens from fossils, flowers, fungi, birds, lizards, mice, butterflies, fishes, flatworms, barnacles, bats and frogs to the parasitical worms in the mouths of rodents and stomachs of birds. The entire range of living things had been captured, examined, dissected, stuffed, pickled, recorded and catalogued. However, these specimens and Darwin's notes were just a personal collection. Only when they were described and published would they become known to the scientific community of the time. Many of his other specimens were described and published in articles by other naturalists. The work of travelling naturalists like Darwin meant that the number of species known in the 1830s–1840s was many times greater than only a century earlier.

Hand-coloured engraving of a Vampire Bat collected by Darwin near Coquimbo, Chile.

delayed completion of the whole set by almost two years through procrastination and ill-health. The 20 plates were drawn by the artist Benjamin Waterhouse Hawkins, who had also illustrated *Fish*, and who later made the giant dinosaur replicas at the Great Exhibition in 1851.

The specimens Darwin described in *Zoology* were deposited in public museums where many are still preserved. The series was originally published in 19 magazine-size numbers with paper covers. When all had at last appeared they could be bound together to form five solid volumes. It was Darwin's most richly illustrated work, with 166 plates, including 82 of coloured birds and mammals. However, the government money ran out before Darwin could write his own volume on marine invertebrates.

The three-volume *Geology* and the five-volume *Zoology*, together with numerous scientific articles and his Journal, were the main publications of Darwin's *Beagle* collections. Yet perhaps his favourite group of living creatures, the marine invertebrates (creatures without backbones), remained unpublished. Darwin at first planned to spend a couple of years publishing descriptions of the most interesting specimens as articles in scientific journals before moving on to his species theory full-time.

THE OFFICIAL NATURALIST

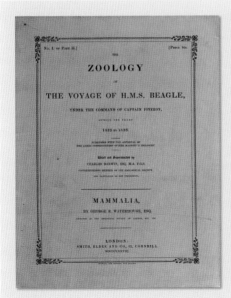

It is sometimes claimed that Darwin was not the "official" naturalist on the *Beagle*. In a sense this is true, he was not an officer in the Royal Navy, but it is a misleading statement. In addition to being formally approved by the Lords of the Admiralty to sail on the *Beagle* for the purpose of being the ship's naturalist, Darwin was always referred to by FitzRoy and others, in letters during the voyage and in the official *Narrative*, as the ship's "naturalist". In addition, the works subsidized and "approved by The Lords Commissioners of Her Majesty's Treasury" were printed with headings like this one from *Zoology*: Edited and Superintended by CHARLES DARWIN, ESQ. M.A., F.R.S., Sec. G.S. NATURALIST TO THE EXPEDITION.

Original paper wrapper or cover to *Mammalia*, by George Robert Waterhouse, one of the 19 numbers of *Zoology of the Voyage of H.M.S. Beagle* edited by Darwin.

76

GEOLOGY OF THE *BEAGLE*

Right at the beginning of the *Beagle* voyage, in the Cape Verde islands, Darwin had the idea of writing a book about the geology of the lands he visited. After returning to England he eventually did publish his work in three volumes between 1842 and 1846.

The first was his theory of the formation of coral reefs and atolls. These curious rings around islands, and the even more mysterious circular-ring islands with their interior lagoons, had long been a mystery to seamen and geologists alike. Some geologists, like Lyell, thought they grew on underwater volcanic craters, but Darwin could not see why so many craters should all stop just below the surface at the shallow depths where coral can live. Already an expert on marine invertebrate life, Darwin applied his geological knowledge of subsidence

to create a brilliant theory. If corals grew around an island, this would be the usual sort of fringing reef. If, however, the sea floor slowly subsided over a long period, then the coral could continue to grow upwards. Eventually, the island could disappear altogether, leaving only a ring-shaped coral atoll.

When the *Beagle* later visited coral atolls, her measurements of the sea's depths and the soundings, which brought up dead broken coral, confirmed that atolls appeared in otherwise very deep waters, far too deep for them to have

Opposite: Glen at Port Desire, drawn by Conrad Martens in his sketchbook on 28 December 1833.

Right: Approach to Montevideo by Conrad Martens, draughtsman on the *Beagle*, published in *Narrative* volume 1 (1839).

DARWIN WAS RIGHT

In 1952 the US Atomic Energy Commission, in preparation for a hydrogen bomb test, drilled into an island of Eniwetok Atoll in the Pacific to discover the distance to the bedrock. If Darwin was wrong there would be a thin layer of coral and then bedrock. The drill passed through 4,158 ft (1,267 meters) of coral before striking volcanic rock – just as Darwin predicted. Contrary to popular belief, the sign "Darwin was right" (opposite, top) was made by Brian Rosen (on the right, John Wells on left), and placed and photographed by him in 1976.

The detonation of a nuclear device during Operation Greenhouse on 8 May 1951 at Eniwetok.

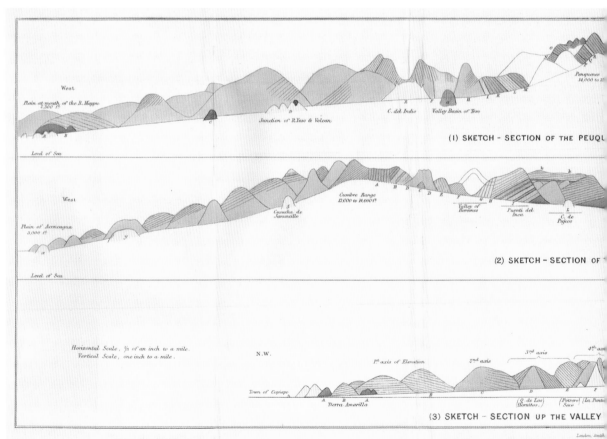

West

Plain at mouth of the R. Mapu
2,500 ft

Junction of R.Yeso & Volcan

C. del Indio   Valley Basin of Yeso

Penquenes
14,000 to 15

(1) SKETCH – SECTION OF THE PEUQU

Level of Sea

West

Plain of Aconcagua
3,000 ft

Cuesta de
Janacillo

Cumbre Range
13,000 to 14,000 ft

Valley of
Horcones

Puenti del
Inca

C. de
Pajos

(2) SKETCH – SECTION OF T

Level of Sea

Horizontal Scale, ⅓ of an inch to a mile.
Vertical Scale, one inch to a mile.

N.W.

1ˢᵗ axis of Elevation   2ⁿᵈ axis   3ʳᵈ axis   4ᵗʰ axis

Town of Copiapo

Tierra Amarilla

(Q. de Los
floritas.

(Potrero)
Seco

(La Punta)

(3) SKETCH – SECTION UP THE VALLEY

London, Smith

Right: The sign erected by Brian Rosen (right of sign) and John Wells (left of sign) when they rediscovered the 1952 borehole in 1976.

Below: Plate 1 from Darwin's Geological observations, second edition (1876), showing geological sections through the Andes. The first two diagrams show his southern and northern traverses between Santiago on the left and Mendoza on the right (circa 200km [124 miles]) and the third shows the Copiapò valley in northern Chile (circa 100km [62 miles]).

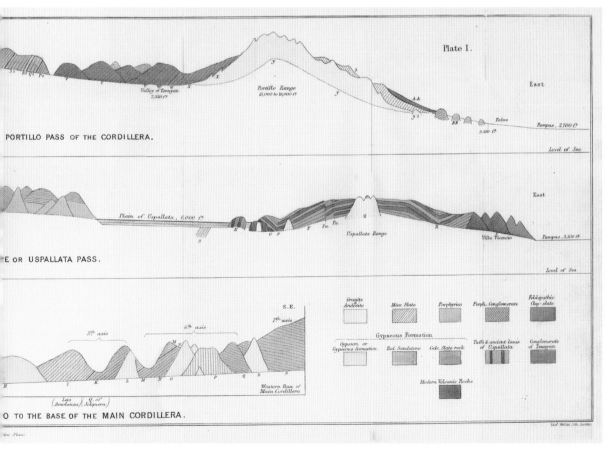

grown from the sea bottom. Darwin consulted many earlier works to compile a map which showed the distribution of the different kinds of reefs. This suggested that large areas of the world's sea floor were rising or subsiding. His theory was instantly well received.

Volcanic Islands, the second book in the series, addressed how volcanic islands came to be formed and shaped. A prevalent earlier theory held that some volcanic islands were shaped like ring craters with steep sides because they grew as a blister-like mass whose top had collapsed inwards making the crater. Darwin argued that the evidence of the rocks showed that, generally, volcanoes were built up by repeated eruptions of lava from the centre, which created an accumulation of volcanic rocks around the vent.

Darwin was convinced that, in general, volcanic areas were regions where the crust of the earth was rising. He contrasted this with areas where the presence of atolls and barrier-reefs indicated subsidence. These vast areas of movement would not be explained until plate tectonics were understood in the 1960s.

In the third book in the series, *Geological Observations on South America*, Darwin demonstrated that there had been great elevations of the earth in parts of South America in very recent geological times. Great earthquakes caused permanent changes in elevation – a process that was still in action. Elsewhere, he observed that rain water, as it percolated through surface rocks, destroyed all traces of shells and other calcareous organisms. This led him to conclude that the

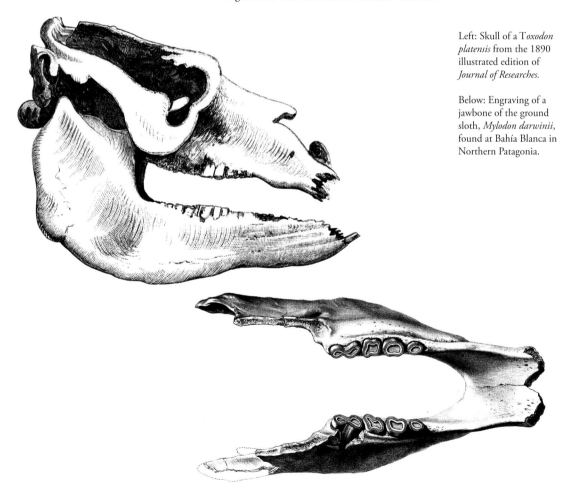

Left: Skull of a T*oxodon platensis* from the 1890 illustrated edition of *Journal of Researches.*

Below: Engraving of a jawbone of the ground sloth, *Mylodon darwinii*, found at Bahía Blanca in Northern Patagonia.

TIME-CONSUMING WORK

Darwin later wrote of his three-volume geological study in his autobiography: "In the early part of 1844, my observations on the Volcanic Islands visited during the voyage of the *Beagle* were published … In 1846 my Geological Observations on South America were published. I record in a little diary, which I have always kept, that my three geological books (Coral Reefs included) consumed four- and-a-half years' steady work; 'and now it is ten years since my return to England. How much time have I lost by illness?'"

A water-coloured drawing by Darwin representing the formation of a coral attol.

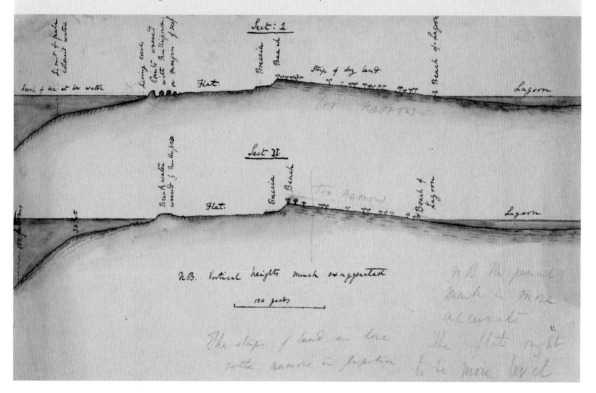

fossil record was subject to such incredible hazards over such long periods that it could never be anything more than isolated fragments of past life.

His account of the vast plains of the Pampas recorded his amazing discoveries of fossil mammals, including *Toxodon, Megatherium, Macrauchenia, Mastodon, Scelidotherium, Megalonyx, Mylodon* and *Glyptodon*. A horse tooth found in old rock beds proved that horses had lived and since become extinct in South America before the Spaniards re-introduced horses in the sixteenth century.

The older view among geologists, that the world possessed universal formations that were the same age everywhere and had the same fossils, was contradicted by Darwin. He found that each great geological age had its own diverse geographical distribution of animals and plants, as the current world does.

Both the theory of coral reefs and the uplifting of the Andes showed how Darwin was learning to solve very large natural puzzles by searching for small mundane causes, which, if reiterated over vast periods of time, could produce enormous changes.

AT HOME WITH THE DARWINS

In September 1842 the Darwins moved to Down House in the sleepy village of Down (now Downe) in Kent, about 25km (15 miles) from the centre of London.

Darwin always wanted to live in a quiet country house. Here he would live and work for the rest of his life with his growing family and several servants, and the countryside around his home became part of his scientific context.

Darwin had very regular habits. He rose early in the mornings and went for an early walk at 7am. He had breakfast at 7:45am before working in his study until 9:30am, his most productive time of the day. Darwin had his study chair elevated on brass legs with castors so he could wheel from desk to table and placed a board over the arms of his chair as a writing surface. At 9:30am he examined his letters in the drawing-room and would lie on the sofa and was read to until 10:30am, when he returned to his study until 12:15pm. He would then go for his walk accompanied by his dog, stopping at his greenhouse to see how his experiments were coming along, before proceeding to the sand-walk, a gravel path around a narrow strip of woodland not far from the house. Then it was time for luncheon. Afterwards he read the newspaper while lying on the sofa, then he wrote his letters. At about 3:00pm he rested on a sofa listening to a novel read by his wife or one of his daughters. At 4.00pm he went for a walk before having coffee and then working again until 5.30pm. Dinner was at 7.30pm, and after dinner he played two games of backgammon

Left: Down House

Above: The old study at Down House as it appears today. The room was restored in 1929 when some of Darwin's children were still alive.

DARWIN'S ILLNESS

For much of life after his marriage Darwin suffered from ill health. We will probably never know exactly what he suffered from, but the consequences are well recorded. He could not bear too much conversation because the excitement caused him to suffer later from vomiting and other symptoms. In the late 1840s he began to take the water cure, an alternative Victorian medicine. He stayed at hydropathic establishments and had an outdoor shower built at home to take ice water ablutions. He felt at first that water cure must be quackery but was later convinced it helped his symptoms.

Daguerrotype of Darwin and his son William taken 23 March 1842. This is the only known photograph of Darwin with a member of his family.

with Emma. They were very competitive and if Emma won Darwin would playfully scold her "bang your bones"! He once wrote to his friend Asa Gray: "Now the tally with my wife in backgammon stands thus: she, poor creature, has won only 2,490 games, whilst I have won, hurrah, hurrah, 2,795 games!" Darwin would then read a scientific book to himself before going to bed at 10:30pm.

He was an affectionate father and kept detailed records of the first words and other mannerisms of his children in a notebook. He was also methodical in recording every penny he spent in account books.

Around 1849 it seems Darwin stopped going to church, though Emma and the children continued to do so. In 1851 his favourite daughter Anne Elizabeth, "Annie", died tragically after a wasting illness. Darwin was devastated but there does not appear to be any evidence for the claim that her death killed off Darwin's Christianity as this had been declining since his return from the *Beagle* voyage in 1836 (see pages 32–33). However, Annie's death was the most distressing event of Darwin's life.

Left: Emma Darwin's 1839 Broadwood grand piano in the drawing room at DownHouse. Emma was a talented pianist and once had lessons from Chopin.

Opposite, bottom left: A photograph of Emma Darwin and her son Leonard, circa 1853.

Opposite, bottom right: One of the garden paths at Down House, through which Darwin would walk.

DARWIN'S CHILDREN

The Darwin family at home around 1863. Left to right Leonard, Etty, Horace, Emma, Bessy, Frank and a friend.

Charles and Emma Darwin had ten children. The first two were born in London and the rest at Down.

William Erasmus (1839–1914) was educated at Christ's College, Cambridge and became a banker in Southampton.

Anne Elizabeth "Annie" (1841–51) was Darwin's favourite daughter who died aged only ten.

Mary Eleanor (1842) lived only 24 days and was thought by Emma to look like her mother, Elizabeth Wedgwood.

Henrietta Emma "Etty" (1843–1927) wrote a life and letters of her mother in 1904, published 1915.

George Howard (1845–1912) who was educated at St John's College, Cambridge, became an astronomer and mathematician.

Elizabeth "Bessy" (1847–1926) never married and little is known about her.

Francis "Frank" (1848–1925) who was educated at Trinity College, Cambridge, became a botanist and wrote Darwin's *Life and Letters* (1887).

Leonard (1850–1943) joined the Royal Engineers in 1871 and taught at the School of Military Engineering.

Horace (1851–1928) became an engineer and founded the Cambridge Scientific Instrument Company in 1885.

Charles Waring (1856–58) died of scarlet fever just before the Darwin–Wallace paper was to be read at the Linnean Society.

Pages from Darwin's "Journal" between 10 May and 31 December 1838. The 29 July trip to Maer to stay with his family is when he discussed marriage with his father.

1838 & reached Shrewsbury July 13th
July 29th Set out for Maer
August 1st London. Began paper
on Glen Roy & finished it
September 6th Finished paper on Glen
Roy — one of the most difficult
& instructive tasks I was ever
employed on
Sept 14th Frittered these days away in working
on Transmutation theories & correcting for ...
Began autumn of alteration theory
Octob 5th Began Coral Paper: requires
much reading
25 Went to Windsor for two days
visit — glorious weather. — delightful.

Very idle at Shrewsbury. some notes
from my Father — & opened 2nd book connected
with metaphysical Enquiries
August. Read a good deal of
various amusing books, & paid
some attention to Metaphysical subjects.
All September read a good deal
on many subjects; thought much upon
religion. — Beginning of October do.

1838
Octob 29th Preface & Addenda on Theory
of Erratic Blocks to Journal
November 9th Started for Maer.
— 11th Sunday. The day of days!
Went to Shrewsbury next day
returned to Maer on the 17th
& to London on the 20th —
December 6th Emma came up to
Town: — most fortunately for me.
December 21st Emma went to Maer
December 31st Entered 12 Upper
Gower St. —

Last ... 7, 8th of November unwell
Wasted entirely the last week
of November. — Beginning of December
prepared number of Birds. From 6th
to 21st busy about House &
sometimes even ones. — To the
end of year. House hunting,
read a little, & wrote
sometimes of being unwell. —

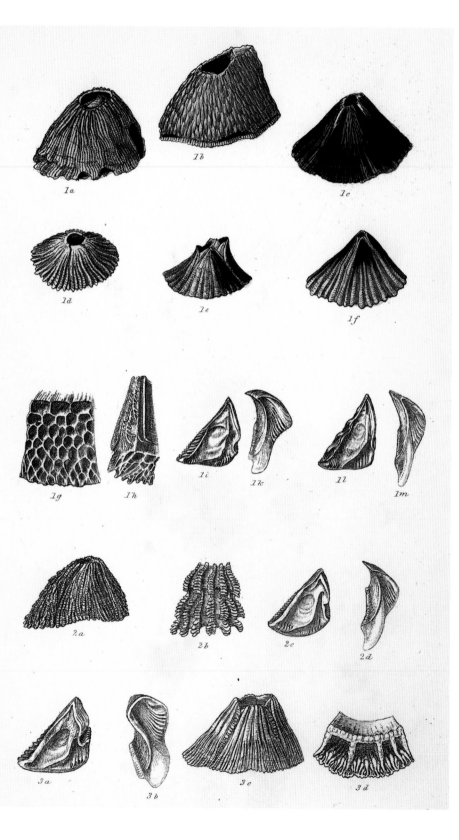

BARNACLES

In late 1846 Darwin had almost completed the decade-long programme of publishing his experiences, theories and collections from the *Beagle* voyage. All that remained were some marine invertebrates.

Opposite: Hand-coloured engraving by George Sowerby showing specimens of the genus *Tetraclita*.

Below: Sketch of parts of a complemental male of *Scalpellum vulgare.*

The government grant which covered *Zoology* had been exhausted, so the plan to write a final volume in that series was dropped. Instead, Darwin planned to spend a couple of years writing up some scientific papers describing his final specimens, and after that he would turn full-time to his species theory. First, he wrote some scientific papers on flatworms and arrow worms. One of his marine invertebrates, collected in the Chonos Archipelago off the west coast of South America, was a curious tiny parasitical barnacle (or cirripede) that bored into the shells of molluscs. It was so different from all other barnacles that Darwin had to name a new sub-order to classify it appropriately. Before long he had made a number of exciting discoveries including a possible insight into the origin of sexual reproduction from hermaphroditic forebears. In order to understand where his barnacles fitted into the vast group of these creatures, he investigated some other genera under his microscope, and was soon encouraged by colleagues to describe the entire sub-class of all known living and fossil barnacles. He thought this might put off his species work for a couple of years, but he was enchanted with his new discoveries, and it was also a pleasure to work with his hands again after only writing books for the past ten years. His health was no longer what it had been in his youth so scientific expeditions remained a thing of the past.

Unfortunately, the barnacle work dragged on and on. Many writers have claimed that Darwin worked at barnacles so long to avoid publishing on evolution or because he felt he needed to bolster his skills or professional reputation. Yet, not only is there no evidence for these claims, but the masses of notes and letters which survive show that Darwin was at first deeply interested and enthusiastic about the barnacle research in its own right. The simple fact of the matter is, as so many scientists and researchers have experienced before and since, that sometimes a

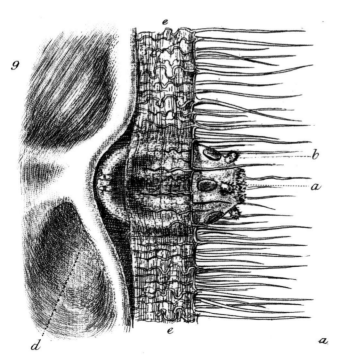

big project can take a lot longer than originally anticipated. For the children growing up in a house where their father had worked for as long as they could remember on barnacles, it seemed everyone did so – one of the children asked a neighbour during a visit "where does he do his barnacles?"

Darwin eventually lost his enthusiasm as the initial exciting discoveries receded into the background. Eventually his once "beloved" barnacles became his "confounded" barnacles. As he wrote to a friend "I hate a Barnacle as no man ever did before, not even a sailor in a slow-sailing ship." His health often failed and, by his own estimate, he lost one to two years during this period to ill health. In the end the barnacle work took eight years to complete. Near the end he would probably have quit early, if he could, but the very nature of making a complete list means one cannot stop halfway through. He had to persevere with it. In the end his work was highly admired by his fellow naturalists, and he was even awarded the prestigious Copley Medal by the Royal Society of London in 1853.

On 9 September 1854 he packed up and returned the last of his borrowed barnacle specimens, noting on the very same day that he "began sorting notes for Species Theory."

Opposite, left: Hand-coloured engraving by George Sowerby showing specimens of the genus *Balanus*.

Opposite, bottom middle: A diagram showing the opthalmic ganglia and eyes of *Lepas fasicularis*.

Opposite, bottom right: Darwin introduced a standardized set of names for the various parts of barnacle shells as illustrated in this woodcut.

CLASSIFYING BARNACLES

Only in the 1830s was it recognized that barnacles are a type of filter-feeding crustacean, not molluscs like clams. Darwin's work discussed the two main types of barnacles – the Balanidae, or sessile (attached by a base rather than a stalk) acorn barnacles, most commonly found attached to rocks; and the Lepadidae, or goose barnacles, commonly found on floating objects. Because barnacles have characters in common with two different crustacean sub-classes, Darwin ranked the barnacles as a separate sub-class of crustacea. Some of the tenets of his species theory were well supported by the barnacle studies, such as the great variability of individuals within a species, the homology of parts in related organisms, the loss of useless organs and the change in function of homologous organs.

Goose barnacles, which are commonly found on floation objects.

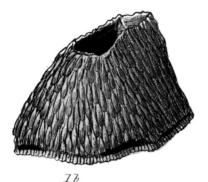

1a

1b

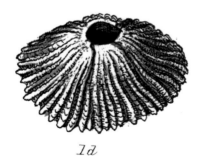

1c

1d

DARWIN'S LENGTHY STUDY OF BARNACLES

Darwin wrote in his autobiography: "… besides describing several new and remarkable forms, I made out the homologies of the various parts – I discovered the cementing apparatus, though I blundered dreadfully about the cement glands – and lastly I proved the existence in certain genera of minute males complemental to and parasitic on the hermaphrodites. … The Cirripedes form a highly varying and difficult group of species to class; and my work was of considerable use to me, when I had to discuss in *On The Origin of Species* the principles of a natural classification." Charles Darwin, *Autobiography*, 1958, page 118.

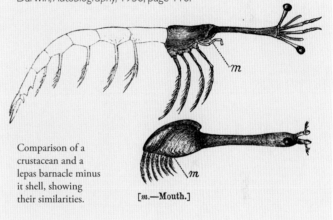

Comparison of a crustacean and a lepas barnacle minus it shell, showing their similarities.

[*m*.—Mouth.]

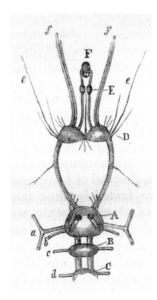

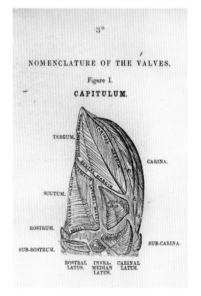

PUTTING THE PUZZLE TOGETHER

After Darwin returned to England following the voyage, occupied with his specimens and books and animated conversations with scientific experts, he took on a tangle of puzzles.

The anatomist Richard Owen confirmed Darwin's suspicions about the South American fossil bones – they belonged to the same kind of creatures, such as armadillos and sloths, that lived uniquely in South America. Ornithologist John Gould told Darwin that his collection of Galápagos grosbeaks, finches and wrens were in fact all finches, and many were island-specific. The mockingbirds, which Darwin had observed at the time to be distinct on different islands, turned out to be separate species, and the same was true for the South American rheas.

Darwin's speculations went through many stops and starts and dead ends. The fact that in all the world the species most similar to those on the Galápagos were on the South American mainland nearly 1,000km (600 miles) to the east was inescapable. That must be their origin.

THE BIRTH OF A THEORY

As Darwin wrote in his autobiography: "In October 1838, that is, fifteen months after I had begun my systematic enquiry, I happened to read for amusement Malthus on Population, and being well prepared to appreciate the struggle for existence which everywhere goes on from long-continued observation of the habits of animals and plants, it at once struck me that under these circumstances favourable variations would tend to be preserved, and unfavourable ones to be destroyed. The result of this would be the formation of new species. Here, then, I had at last got a theory by which to work."

The political economist Thomas Malthus whose stress on the inevitably high rates of population growth inspired Darwin.

Below: The ornithologist John Gould told Darwin that the varieties of similar birds he collected on the Galápagos were in fact distinct species.

It was inexplicable on any other view – the climate, temperature the very bedrock of the Galápagos was different from South America. All of this evidence converged on the general conclusion that species had to be changeable.

Charles Lyell argued that species in the fossil record had naturally become extinct as the world changed so that it no longer suited them. New species were somehow created to suit the new environment, and from a centre of creation they would have migrated outwards. But if, as Darwin then believed, his finches all fed together in flocks, how could the same environment bring about different species?

He started at the most basic level with reproduction, considering the question of why organisms reproduce, and why they have such short life spans and don't live forever. He considered the two kinds of reproduction – splitting or budding, which resulted in identical copies, and sexual reproduction, which resulted in mixed and, therefore, altered offspring.

Given that the world changed radically over time, reproduction which resulted in more variable offspring would allow them "to adapt & alter the race to changing world". If species were descended from earlier species, just as an individual is descended from parents, then different species in a genus would be related by common descent, as cousins share the same grandparents. He sketched a tiny tree diagram to demonstrate how species would thus be related by lineage.

It was not until 28 September 1838 that he read Thomas Malthus's *Essay on the Principle of Population* (1826). Malthus argued that human population growth, unless somehow checked, would outstrip food production. Population growth, according to Malthus, should be geometrical. For example, two parents might have four children, each of whom could have four children, whose children could also have four children and so forth.

The focus of this argument instantly inspired Darwin. He realized that an enormous proportion of the living things produced were always destroyed before they could reproduce. This must have been true because every species would otherwise have bred enough to cover the Earth. Instead, populations remained roughly stable year after year. The only way this could be so was if most offspring (from pollen, to seeds and eggs) did not survive long enough to reproduce.

Darwin, already concentrating on how new varieties of life might be formed, suddenly realized that the key was whatever made a difference between those that survived to reproduce and those that did not. He later came to call this open-ended collection of causes "natural selection" because it was analogous to breeders choosing which individuals to breed from and thus changing a breed markedly over time.

Darwin imagined the convulsing cosmos of living things all over the globe, all reproducing at a fantastic rate, and almost all being ruthlessly destroyed, devoured, starved or lost. The bursting outward force of reproduction was checked by the carnage of ingestion and death.

These two opposing processes were like a war of nature that never ended. Yet those with the right stuff to survive would pass on their characteristics to offspring. The result would be the changing of species over time and, most crucially, the way by which they could become adapted to particular environments.

Every part of every organism varies. There has always been an endless and spontaneous supply of variety. If circumstances arose which meant that one of these variations happened to benefit its possessor, then that variety would survive the filtering process of natural selection and be passed on. In this simple and natural manner, every change from the unknown ancestor of *Glyptodon* to a modern armadillo could be effected.

Quietly, without any fanfare, a young man sitting in his London study surrounded by boxes of specimens and scratching his nib across the page had grasped, for the first time, how all living things are related. It would change the world forever.

Below: A dwarf armadillo collected by Darwin at Bahia Blanca.

THE PICHI ARMADILLO,
DASYPUS MINUTUS.
Buenos Aires. Zaedyus pichiy
Presented by Charles Darwin, Esq., 1855.

DARWIN'S DELAY

For many years it was believed that Darwin kept his theory a secret because he was afraid of what his colleagues would think. Most cite the famous letter in which Darwin wrote "species are not (it is like confessing a murder) immutable". Yet this was typical Darwin humour. He once wrote to a friend "When I saw your bundle of observations, I felt as if I had committed theft, arson or murder." The fact is Darwin discussed his belief in evolution with many friends and colleagues. It was not fear that delayed the publication of his theory but a complex cocktail of finishing earlier projects, spiralling barnacles studies and a massive research programme that took many years to complete.

A 27 March 1855 letter from Darwin to his cousin Fox in which he tells of his new project – species.

Opposite: Extracts from Darwin's first sketch of his species theory. As he recalled "In June 1842 I first allowed myself the satisfaction of writing a very brief abstract of my theory in pencil in 35 pages".

Below: Darwin's sketch in his Notebook B (1837) representing diverging descendant species derived from a common ancestor (1). The lines that end without a cross bar represent extinct species.

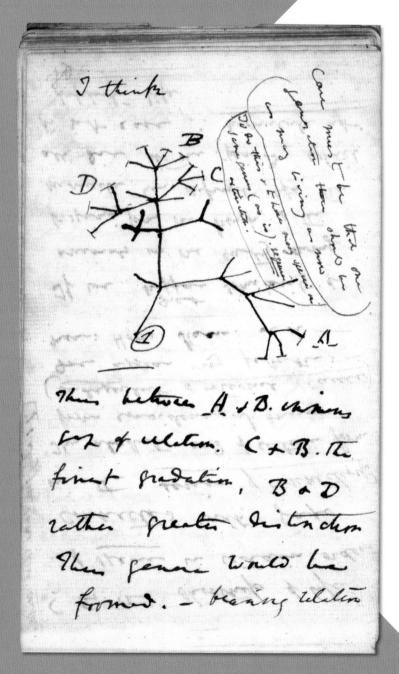

ON THE ORIGIN OF SPECIES

For almost two years Darwin worked on his species theory on a vast scale. In May 1856 Charles Lyell had warned him that other naturalists were drawing closer to Darwin's views and suggested that he should condense his materials and bring out a book more quickly.

Darwin followed this advice. From mid-1856 he worked steadily on one chapter after another. These included chapters on geographical distribution, variation in nature, the struggle for existence, natural selection, hybridism, divergence, instinct and so forth. He worked diligently, though often interrupted by his poor health, which had plagued him increasingly since his return from the *Beagle* voyage.

By the summer of 1858 Darwin was about halfway through this "big book" on species which he planned to call *Natural Selection*. One fateful morning, however, a packet arrived from an English naturalist and collector in Indonesia, named Alfred Russel Wallace, with whom Darwin had been corresponding. Darwin was thunderstruck to see that, in an enclosed essay, Wallace proposed a theory for the origin of species strikingly similar to his own. Darwin

feared his priority would be lost, but unselfishly forwarded the essay to Lyell as Wallace had requested. Rather than see their friend lose his priority of 20 years, Lyell, in consultation with Joseph Dalton Hooker, arranged to have papers by both Wallace and Darwin read together at a meeting of the Linnean Society of London in July 1858. When later printed, it was the first publication of the theory of evolution by natural selection; however, the paper made surprisingly little impact at first.

Darwin was urged to publish a summary or abstract of his large unfinished work, which he spent the next 13 months drafting. This summary became *On the Origin of Species by Means of Natural Selection*. One would never know it from the modest way he spoke of it in his letters, but the book would change the world forever. Its importance for our understanding of life on Earth is difficult to exaggerate. Between the covers of a single

Opposite: Darwin's Old Study in Down House, Kent. *On the Origin of Species* was written in the chair on a board.

Right: Alfred Russel Wallace in 1848.

book Darwin managed to demonstrate the most fundamental patterns of past and present life on Earth, and at a stroke all of the hundreds of families, genera and countless thousands of species were connected in one single and beautifully simple system. All of life is related, genealogically, on a great branching tree, the tree of life. He called it "the theory of descent with modification through natural selection".

The book had 14 chapters plus an introduction and conclusion. Darwin began by explaining how he came to doubt the stability of species and how long he had worked on the subject. The brute facts of the similarities of different species, the similarities during embryological development of members of the same genus, geographical distribution, the progressive succession of fossil forms and so forth could indeed convince a naturalist that species change, but this would still be incomplete. One would need a theory to explain how they changed and, most importantly of all, how they came to be so wonderfully adapted to their environments and their immensely complex relationships with one another. Darwin's theory of natural selection explained how adaptation could occur, over many generations, given the commonly accepted, but often overlooked, properties of living things.

The book was about how all species originate, not about the origin of any one species, and he did make it very clear that mankind was part of nature: "Light will be thrown on the origin of man and his history."

Darwin felt that *On the Origin of Species* was "no doubt the chief work of my life". The first edition sold out to the bookseller trade (not the public) on the first day.

Oppsoite top: An artist rendition of Darwin in his study. Oil painting by E. Eustafieff, 1958.

Opposite right: *On the Origin of Species* published in November 1859 by John Murray.

OBSERVING THE TRANSMUTATION OF SPECIES

"From September 1854 onwards I devoted all my time to arranging my huge pile of notes, to observing, and experimenting, in relation to the transmutation of species. During the voyage of the *Beagle* I had been deeply impressed by discovering in the Pampean formation great fossil animals covered with armour like that on the existing armadillos; secondly, by the manner in which closely allied animals replace one another in proceeding southwards over the Continent; and thirdly, by the South American character of most of the productions of the Galápagos archipelago, and more especially by the manner in which they differ slightly on each island of the group; none of these islands appearing to be very ancient in a geological sense. It was evident that such facts as these, as well as many others, could be explained on the supposition that species gradually become modified; and the subject haunted me." Charles Darwin, *Autobiography*, 1958, pages 118–119.

A diagram showing the divergence of daughter species from ancestral forms. This is the only illustration in *On the Origin of Species*.

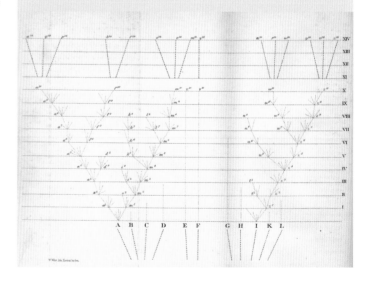

ENDLESS FORMS MOST BEAUTIFUL

"… all living things have much in common, in their chemical composition, their germinal vesicles [eggs and sperm], their cellular structure, and their laws of growth and reproduction. … There is grandeur in this view of life, with its several powers, having been originally breathed into a few forms or into one; and that, whilst this planet has gone cycling on according to the fixed law of gravity, from so simple a beginning endless forms most beautiful and most wonderful have been, and are being, evolved."
Charles Darwin, *On the Origin of Species*, 1859, page 484

The title page of the first edition of *On the Origin of Species* (1859) "the chief work of my life" as Darwin recalled in his autobiography.

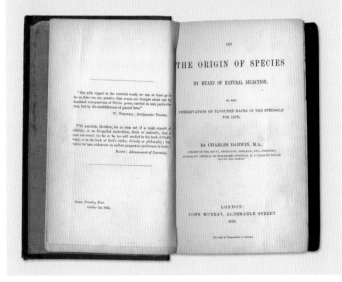

JOURNAL

⌘

THE PROCEEDINGS

⌘

THE LINNEAN SOCIETY.

ZOOLOGY.

VOL III

LONDON
LONGMAN, BROWN, GREEN, LONGMAN & ROBERTS

AND

WILLIAMS AND NORGATE

1858

Drafts of Darwin's unpublished "big book" on species from 1857. These extracts discuss geographical distribution and representative species.

On the Tendency of Species to form Varieties; and on the Perpetuation of Varieties and Species by Natural Means of Selection. By CHARLES DARWIN, Esq., F.R.S., F.L.S., & F.G.S., and ALFRED WALLACE, Esq. Communicated by Sir CHARLES LYELL, F.R.S., F.L.S., and J. D. HOOKER, Esq., M.D., V.P.R.S., F.L.S., &c.

[Read July 1st, 1858.]

London, June 30th, 1858.

MY DEAR SIR,—The accompanying papers, which we have the honour of communicating to the Linnean Society, and which all relate to the same subject, viz. the Laws which affect the Production of Varieties, Races, and Species, contain the results of the investigations of two indefatigable naturalists, Mr. Charles Darwin and Mr. Alfred Wallace.

These gentlemen having, independently and unknown to one another, conceived the same very ingenious theory to account for the appearance and perpetuation of varieties and of specific forms on our planet, may both fairly claim the merit of being original thinkers in this important line of inquiry; but neither of them having published his views, though Mr. Darwin has for many years past been repeatedly urged by us to do so, and both authors having now unreservedly placed their papers in our hands, we think it would best promote the interests of science that a selection from them should be laid before the Linnean Society.

Taken in the order of their dates, they consist of:—

1. Extracts from a MS. work on Species*, by Mr. Darwin, which was sketched in 1839, and copied in 1844, when the copy was read by Dr. Hooker, and its contents afterwards communicated to Sir Charles Lyell. The first Part is devoted to "The Variation of Organic Beings under Domestication and in their Natural State;" and the second chapter of that Part, from which we propose to read to the Society the extracts referred to, is headed, "On the Variation of Organic Beings in a state of Nature; on the Natural Means of Selection; on the Comparison of Domestic Races and true Species."

2. An abstract of a private letter addressed to Professor Asa Gray, of Boston, U.S., in October 1857, by Mr. Darwin, in which

* This MS. work was never intended for publication, and therefore was not written with care.—C. D. 1858.

he repeats his views, and which shows that these remained unaltered from 1839 to 1857.

3. An Essay by Mr. Wallace, entitled "On the Tendency of Varieties to depart indefinitely from the Original Type." This was written at Ternate in February 1858, for the perusal of his friend and correspondent Mr. Darwin, and sent to him with the expressed wish that it should be forwarded to Sir Charles Lyell, if Mr. Darwin thought it sufficiently novel and interesting. So highly did Mr. Darwin appreciate the value of the views therein set forth, that he proposed, in a letter to Sir Charles Lyell, to obtain Mr. Wallace's consent to allow the Essay to be published as soon as possible. Of this step we highly approved, provided Mr. Darwin did not withhold from the public, as he was strongly inclined to do (in favour of Mr. Wallace), the memoir which he had himself written on the same subject, and which, as before stated, one of us had perused in 1844, and the contents of which we had both of us been privy to for many years. On representing this to Mr. Darwin, he gave us permission to make what use we thought proper of his memoir, &c.; and in adopting our present course, of presenting it to the Linnean Society, we have explained to him that we are not solely considering the relative claims to priority of himself and his friend, but the interests of science generally; for we feel it to be desirable that views founded on a wide deduction from facts, and matured by years of reflection, should constitute at once a goal from which others may start, and that, while the scientific world is waiting for the appearance of Mr. Darwin's complete work, some of the leading results of his labours, as well as those of his able correspondent, should together be laid before the public.

We have the honour to be yours very obediently,
CHARLES LYELL.
JOS. D. HOOKER.

J. J. Bennett, Esq.,
Secretary of the Linnean Society.

1. *Extract from an unpublished Work on Species, by C. DARWIN, Esq., consisting of a portion of a Chapter entitled, "On the Variation of Organic Beings in a state of Nature; on the Natural Means of Selection; on the Comparison of Domestic Races and true Species."*

De Candolle, in an eloquent passage, has declared that all nature is at war, one organism with another, or with external nature.

Seeing the contented face of nature, this may at first well be doubted; but reflection will inevitably prove it to be true. The war, however, is not constant, but recurrent in a slight degree at short periods, and more severely at occasional more distant periods; and hence its effects are easily overlooked. It is the doctrine of Malthus applied in most cases with tenfold force. As in every climate there are seasons, for each of its inhabitants, of greater and less abundance, so all annually breed; and the moral restraint which in some small degree checks the increase of mankind is entirely lost. Even slow-breeding mankind has doubled in twenty-five years; and if he could increase his food with greater ease, he would double in less time. But for animals without artificial means, the amount of food for each species must, *on an average*, be constant, whereas the increase of all organisms tends to be geometrical, and in a vast majority of cases at an enormous ratio. Suppose in a certain spot there are eight pairs of birds, and that *only* four pairs of them annually (including double hatches) rear only four young, and that these go on rearing their young at the same rate, then at the end of seven years (a short life, excluding violent deaths, for any bird) there will be 2048 birds, instead of the original sixteen. As this increase is quite impossible, we must conclude either that birds do not rear nearly half their young, or that the average life of a bird is, from accident, not nearly seven years. Both checks probably concur. The same kind of calculation applied to all plants and animals affords results more or less striking, but in very few instances more striking than in man.

Many practical illustrations of this rapid tendency to increase are on record, among which, during peculiar seasons, are the extraordinary numbers of certain animals; for instance, during the years 1826 to 1828, in La Plata, when from drought some millions of cattle perished, the whole country actually *swarmed* with mice. Now I think it cannot be doubted that during the breeding-season all the mice (with the exception of a few males or females in excess) ordinarily pair, and therefore that this astounding increase during three years must be attributed to a greater number than usual surviving the first year, and then breeding, and so on till the third year, when their numbers were brought down to their usual limits on the return of wet weather. Where man has introduced plants and animals into a new and favourable country, there are many accounts in how surprisingly few years the whole country has become stocked with them. This increase would

necessarily stop as soon as the country was fully stocked; and yet we have every reason to believe, from what is known of wild animals, that *all* would pair in the spring. In the majority of cases it is most difficult to imagine where the checks fall—though generally, no doubt, on the seeds, eggs, and young; but when we remember how impossible, even in mankind (so much better known than any other animal), it is to infer from repeated casual observations what the average duration of life is, or to discover the different percentage of deaths to births in different countries, we ought to feel no surprise at our being unable to discover where the check falls in any animal or plant. It should always be remembered, that in most cases the checks are recurrent yearly in a small, regular degree, and in an extreme degree during unusually cold, hot, dry, or wet years, according to the constitution of the being in question. Lighten any check in the least degree, and the geometrical powers of increase in every organism will almost instantly increase the average number of the favoured species. Nature may be compared to a surface on which rest ten thousand sharp wedges touching each other and driven inwards by incessant blows. Fully to realize these views much reflection is requisite. Malthus on man should be studied; and all such cases as those of the mice in La Plata, of the cattle and horses when first turned out in South America, of the birds by our calculation, &c., should be well considered. Reflect on the enormous multiplying power *inherent and annually in action* in all animals; reflect on the countless seeds scattered by a hundred ingenious contrivances, year after year, over the whole face of the land; and yet we have every reason to suppose that the average percentage of each of the inhabitants of a country usually remains constant. Finally, let it be borne in mind that this average number of individuals (the external conditions remaining the same) in each country is kept up by recurrent struggles against other species or against external nature (as on the borders of the Arctic regions, where the cold checks life), and that ordinarily each individual of every species holds its place, either by its own struggle and capacity of acquiring nourishment in some period of its life, from the egg upwards; or by the struggle of its parents (in short-lived organisms, when the main check occurs at longer intervals) with other individuals of the *same* or *different* species.

But let the external conditions of a country alter. If in a small degree, the relative proportions of the inhabitants will in most cases simply be slightly changed; but let the number of inhabitants be small, as on an island, and free access to it from other countries be circumscribed, and let the change of conditions continue progressing (forming new stations), in such a case the original inhabitants must cease to be as perfectly adapted to the changed conditions as they were originally. It has been shown in a former part of this work, that such changes of external conditions would, from their acting on the reproductive system, probably cause the organization of those beings which were most affected to become, as under domestication, plastic. Now, can it be doubted, from the struggle each individual has to obtain subsistence, that any minute variation in structure, habits, or instincts, adapting that individual better to the new conditions, would tell upon its vigour and health? In the struggle it would have a better *chance* of surviving; and those of its offspring which inherited the variation, be it ever so slight, would also have a better *chance*. Yearly more are bred than can survive; the smallest grain in the balance, in the long run, must tell on which death shall fall, and which shall survive. Let this work of selection on the one hand, and death on the other, go on for a thousand generations, who will pretend to affirm that it would produce no effect, when we remember what, in a few years, Bakewell effected in cattle, and Western in sheep, by this identical principle of selection?

To give an imaginary example from changes in progress on an island:—let the organization of a canine animal which preyed chiefly on rabbits, but sometimes on hares, become slightly plastic; let these same changes cause the number of rabbits very slowly to decrease, and the number of hares to increase; the effect of this would be that the fox or dog would be driven to try to catch more hares: his organization, however, being slightly plastic, those individuals with the lightest forms, longest limbs, and best eyesight, let the difference be ever so small, would be slightly favoured, and would tend to live longer, and to survive during that time of the year when food was scarcest; they would also rear more young, which would tend to inherit these slight peculiarities. The less fleet ones would be rigidly destroyed. I can see no more reason to doubt that these causes in a thousand generations would produce a marked effect, and adapt the form of the fox or dog to the catching of hares instead of rabbits, than that greyhounds can be improved by selection and careful breeding. So would it be with plants under similar circumstances. If the number of individuals of a species with plumed seeds could be increased by greater powers of dissemination within its own area

(that is, if the check to increase fell chiefly on the seeds), those seeds which were provided with ever so little more down, would in the long run be most disseminated; hence a greater number of seeds thus formed would germinate, and would tend to produce plants inheriting the slightly better-adapted down[*].

Besides this natural means of selection, by which those individuals are preserved, whether in their egg, or larval, or mature state, which are best adapted to the place they fill in nature, there is a second agency at work in most unisexual animals, tending to produce the same effect, namely, the struggle of the males for the females. These struggles are generally decided by the law of battle, but in the case of birds, apparently, by the charms of their song, by their beauty or their power of courtship, as in the dancing rock-thrush of Guiana. The most vigorous and healthy males, implying perfect adaptation, must generally gain the victory in their contests. This kind of selection, however, is less rigorous than the other; it does not require the death of the less successful, but gives to them fewer descendants. The struggle falls, moreover, at a time of year when food is generally abundant, and perhaps the effect chiefly produced would be the modification of the secondary sexual characters, which are not related to the power of obtaining food, or to defence from enemies, but to fighting with or rivalling other males. The result of this struggle amongst the males may be compared in some respects to that produced by those agriculturists who pay less attention to the careful selection of all their young animals, and more to the occasional use of a choice mate.

II. *Abstract of a Letter from* C. DARWIN, Esq., *to* Prof. ASA GRAY, *Boston, U.S., dated Down, September 5th, 1857.*

1. It is wonderful what the principle of selection by man, that is the picking out of individuals with any desired quality, and breeding from them, and again picking out, can do. Even breeders have been astounded at their own results. They can act on differences inappreciable to an uneducated eye. Selection has been *methodically* followed in *Europe* for only the last half century; but it was occasionally, and even in some degree methodically, followed in the most ancient times. There must have been also a kind of unconscious selection from a remote period, namely in

[*] I can see no more difficulty in this, than in the planter improving his varieties of the cotton plant.—C. D. 1858.

N° 10

LA Petite LUNE

Bureaux : rue Coq-Héron, 5 || **Dessins de GILL** || Abonnem¹ˢ : Paris, 3 fr. — Dépar¹ˢ, 3 fr. 50

5 C^mes

DARWIN

ARBRE DE LA SCIENCE

(Voir à la page 2.)

THE RECEPTION OF DARWIN'S THEORY OF EVOLUTION

Darwin's great work *On the Origin of Species by Means of Natural Selection, or the Preservation of Favoured Races in the Struggle for Life* was published in November 1859. It is often said that a great controversy of science and religion erupted.

Opposite: The cover of Parisian satirical journal *La Petite Lune* (circa 1878), featuring a caricature of Darwin as a monkey hanging from the tree of science.

Below: A French cartoon from *La Petite Lune* depicting Darwin as a monkey smashing through circus hoops labelled Credulity, Superstition, Errors and Ignorance.

This is very far from the case. In fact, most of the heat had been spent over radically naturalistic works published in the preceding decades.

Books like George Combe's *Constitution of Man* (1828), and the anonymous *Vestiges of Creation* (1844) had shocked readers far more with their theories about natural laws controlling all of the universe, mankind included, and had also been read by larger audiences. These earlier works, after all, had societies founded to oppose them and, in at least one case, were even publicly burned. No such treatment awaited *On the Origin of Species* or its author.

Nevertheless, *On the Origin of Species* sparked off a world-wide debate. There were hundreds of book reviews and countless works written in opposition or in support, and a second edition of 3,000 copies was printed in January 1860.

Darwin was not particularly interested in the reception of his theory by the general public. He was mostly concerned with the views of scientists with the relevant knowledge that was needed to assess his work. At first, the views were mixed. Darwin had expected religious and ideological objections, but he was surprised by the accusations that his methods were "unscientific". Sedgwick, for example, complained that Darwin had "deserted the true inductive track", and wrote to Darwin "I have read your book with more pain than pleasure. Parts of it I admired greatly, parts I laughed at till my sides were almost sore; other parts I read with absolute sorrow, because I think

THE CLASH OF SCIENCE AND RELIGION?

Darwin's theory was debated before a large audience at the 1860 meeting of the British Association for the Advancement of Science in Oxford. What happened there has become one of the main myths in the Darwinian story. Supposedly, a raw clash between science and religion was manifested when Bishop Samuel Wilberforce asked the plucky young naturalist Thomas Henry Huxley whether he was descended from an ape on his grandmother's or grandfather's side. Huxley, again according to legend, replied that he would prefer an ape than a man who used his gifts to bring ridicule into serious scientific discussion. In fact, no matter what was really said, the debate was more about personalities and egos than science versus religion.

Bishop Samuel Wilberforce.

them utterly false and grievously mischievous." Darwin wrote to Lyell, "I have heard, by a roundabout channel, that Herschel says my book 'is the law of higgeldy-piggelty.' What this exactly means I do not know, but it is evidently very contemptuous. If true this is a great blow and discouragement."

Nevertheless, Darwin, as a prominent and respected name in science, had to be taken seriously. Different kinds of people reacted differently to various components and implications of his theory. Many scientists, especially younger ones, soon accepted that evolution was true. The phrenologist and botanist Hewett Cottrell Watson wrote to Darwin "Your leading idea will assuredly become recognised as an established truth in science, i.e. 'Natural selection.' It has the characteristics of all great natural truths, clarifying what was obscure, simplifying what was intricate, adding greatly to previous knowledge. You are the greatest revolutionist in natural history of this century, if not of all centuries." Some, however, did not accept Darwin's stress on natural selection. Very many writers focused on the implication that humans must be descended from earlier species. For many, especially religious and non-scientific readers, this was considered unacceptable and Darwin was sometimes harshly criticized.

Right: The London *Archaeopteryx* specimen. According to Darwin "Hardly any recent discovery shows more forcibly than this how little we as yet know of the former inhabitants of the world."

In 1861 Henry Walter Bates, a naturalist just back from Brazil, showed that natural selection could explain the mystery of mimicry in South American butterflies. Bates had found that many types of brightly coloured butterflies escaped being eaten because they had a very unpleasant odour. Wherever such species existed, different rarer butterfly species, even from different families, had evolved to look strikingly like them. The more one of these mimics resembled one of the inedible species, the greater its chances of being left alone by birds.

More and more scientists found that Darwin's explanation made sense of their particular areas of expertise. Articles and books began to appear praising Darwin's ideas, and by around 1869, ten years after the first publication, most scientists had accepted that Darwin was right. Of course, things were not the same everywhere. In Germany the theory was accepted rather quickly and with little fuss, whereas in France it was ignored and scorned for many years. But by the 1870s Darwin was widely regarded as a scientific revolutionary who had transformed the study of the natural world.

Opposite: Plate depicting mimicry of brightly coloured butterflies ignored by birds from "On the lepidoptera of the Amazon Valley" *Transactions of the Linnean Society* by H W Bates, 1862.

A NEW DIRECTION

"The history of every science shows that the great epochs of its progress are those not so much of new discoveries of facts, as of those new ideas which have served for the colligation of facts previously known into general principles, and which have thenceforward given a new direction to inquiry. … Naturalists have gone on quite long enough on the doctrine of the 'permanence of species.' … the difficulty of distinguishing between true species and varieties increases, instead of diminishing … The doctrine of progressive modification by Natural Selection propounded by Mr. Darwin, will give a new direction to inquiry into the real genetic relationship of species, existing and extinct…" The Physician William Benjamin Carpenter reviewing *On the Origin of Species* in 1860.

The American palaeontologist O C Marsh discovered a series of equine fossils in the late 1860s and 1870s. The collection was the first to demonstrate the line of descent of an existing animal. This engraving shows the loss of toes as the animals became larger and adapted for grazing and running.

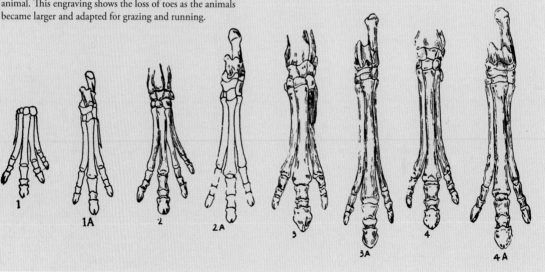

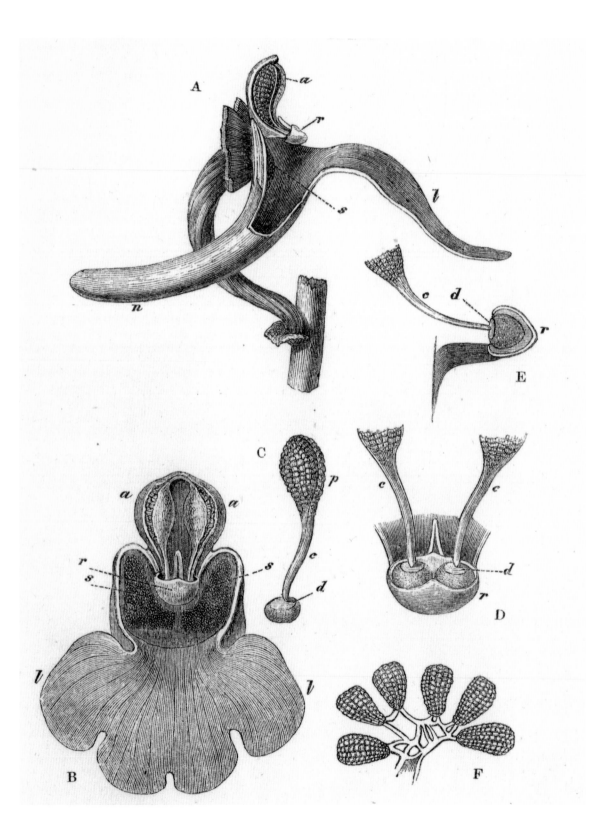

ORCHIDS

It may seem surprising, but Darwin's next book was on the pollination mechanisms of orchids, with which he had become fascinated almost by accident. Since 1859 he had attended to three English and two American editions of *On the Origin of Species* and two translations, and the book on orchids was Darwin's first detailed demonstration of the power of natural selection.

Linnaeus had known of 100 species of orchids, but by 1860 there were 433 known genera with 6,000 species. At the time, rare and beautiful orchids were all the rage among wealthy collectors, and Darwin prevailed on many of them to send orchids to him.

Critics of Darwin's theory of evolution by natural selection had claimed that natural selection could not explain the details of living things. The delicate curves and structures of beautiful flowers was one example. Some believed they were created for their beauty to human eyes, but Darwin knew this could not be true since some flowers appeared only at night and others in parts of the world uninhabited

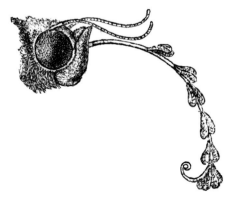

Opposite: Diagram showing the relative position of parts of the flower of *Orchis mascula* with special details of the pollen-mass or pollinium.

Right: A moth found with its proboscis covered in perfectly placed pollen packets from orchids. The mystery of how orchids were pollinated was solved.

by human beings. Part of Darwin's great and growing enthusiasm for orchids came from his discovery that natural selection could explain the most apparently trivial curves and shapes of tiny blossoms.

Darwin experimented with orchids to find out how they were pollinated. He tried pollinating them artificially with their own pollen, but these almost never produced fertile seeds. He covered one plant with a glass dome and left others uncovered beside it. With each day Darwin recorded that the pollen-carrying parcels, called pollinia, gradually disappeared from the uncovered plants, whereas the covered plant's pollinia remained undisturbed. In time, the uncovered plants produced fertile seed, while the covered plant produced none.

With the help of other naturalists, Darwin found that 23 species of moths had been found with orchid pollinia attached to their probosces. He showed that pollinia were adapted for removal by insects as they probed the flower's depths for its nectar. As an insect probed for nectar, the pollinia became glued to the insect and detached from the flower as the insect withdrew and flew away. The position of the firmly attached pollinia on the insect was also just right to fertilize the next orchid that was visited by the insect when it again probed

PLANT REPRODUCTION

The fact that plants and flowers have different sexes had only been established around 60 years before Darwin began his work on orchids. The male organ, the anther, produces pollen (the yellow powder that stains our fingers), which must reach the pistil or female organ in order for the plant to produce fertile seeds. It is often possible to fertilize artificially (or pollinate) a flower by transferring pollen with a small paint brush. How the pollen is transferred varies. Sometimes it is simply blown by the wind from one plant to another, but in most orchids the pollen is so firmly embedded that it cannot fall out. This was a mystery to Darwin. How could orchids be pollinated? As no one had ever observed insects pollinating orchids, some people speculated that orchids simply pollinated themselves.

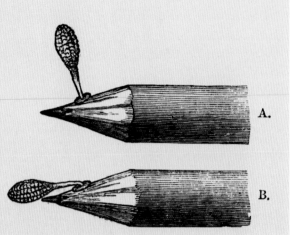

A. Pollen-mass of *Orchis mascula*, when first attached.
B. Pollen-mass of *Orchis mascula* after the act of depression.

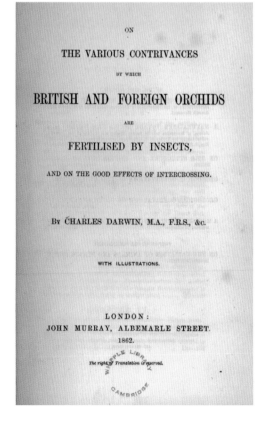

ON

THE VARIOUS CONTRIVANCES

BY WHICH

BRITISH AND FOREIGN ORCHIDS

ARE

FERTILISED BY INSECTS,

AND ON THE GOOD EFFECTS OF INTERCROSSING.

BY CHARLES DARWIN, M.A., F.R.S., &c.

WITH ILLUSTRATIONS.

LONDON:
JOHN MURRAY, ALBEMARLE STREET.
1862.

The right of Translation is reserved.

ASA GRAY (1810–88)

The American botanist Asa Gray praised Darwin in 1874 for providing "the explanation of all these and other extraordinary structures, as well as of the arrangement of blossoms in general, and even the very meaning and need of sexual propagation … The aphorism 'Nature abhors close fertilization,' and the demonstration of the principle, belong to our age, and to Mr. Darwin. To have originated this, and also the principle of Natural Selection – the truthfulness and importance of which are evident the moment it is apprehended – and to have applied these principles to the system of nature in such a manner as to make, within a dozen years, a deeper impression upon natural history than has been made since Linnaeus, are ample title for one man's fame." Notice by Asa Gray in *The American Naturalist* 1874, 8, Number 8. (August): 475–479, page 478.

for a drink of nectar. Darwin demonstrated that the flowers had just the right structure to guide insects into the right position to cause the pollinias to stick in this way and to receive pollen from visiting insects. In this way, he demonstrated the complex interdependence of living things.

Darwin determined that some small orchids bear between 6,000 and 186,000 seeds per plant. He then calculated that if one plant's seeds successfully grew, they would cover an acre, and that at the same rate of increase the great-grandchildren of the plant would cover all the land on earth. Clearly this did not happen, which meant that severe checks prevented orchids from reproducing to this extent. Darwin was not sure what these checks might be, but he did conclude that any naturally occurring differences that allowed a plant to reproduce successfully were likely to proliferate.

Darwin showed that complex adaptations ensured that flowers would be fertilized by different individual plants. This confirmed his speculation that long continued self-reproduction, as with a single hermaphrodite, would lose out to a sexually reproducing creature – because each new generation was a new unique mixture and thus offered more variability to survive the rigours of the struggle for existence. He showed that plants were just as rich with intricate adaptations as animals, and that their adaptations were cobbled together from pre-existing elements by natural selection. Darwin was thrilled to discover the naturalistic explanation for such beautiful structures.

Opposite, left: The cover of *Orchids* decorated with a gilt orchid. The book was published on 15 May 1862.

Opposite, right: The title page of *Orchids*, Darwin's first book after *On the Origin of Species*, which showed the detailed relationships between the sexual structures of orchids and the insects which pollinate them.

VARIATION

After publication of *On the Origin of Species* Darwin did not abandon his intention of writing his large work on evolution. The first two chapters of his unpublished manuscript became *Variation of Animals and Plants under Domestication*, published in 1868.

Opposite: A brown and white English Pouter pigeon. Breeders can shape animals to extraordinary degrees by selective breeding.

Below: Side views of pigeon skulls from four different breeds, which show how radically a single species could vary.

It is his longest book, and at that time he envisioned writing two more works of equal scope, although he never did.

At the beginning of the work Darwin described the kinds of evidence that convinced him of evolution:

To exhume with one's own hands the bones of extinct and gigantic quadrupeds brings the whole question of the succession of species vividly before one's mind; and I had found in South America great pieces of tesselated armour exactly like, but on a magnificent scale, that covering the pigmy armadillo; I had found great teeth like those of the living sloth, and bones like those of the cavy. An analogous succession of allied forms had been previously observed in Australia. Here then we see the prevalence, as if by descent, in time as in space, of the same types in the same areas; and in neither case does the similarity of the conditions by any means seem sufficient to account for the similarity of the forms of life. It is notorious that the fossil remains of closely consecutive formations

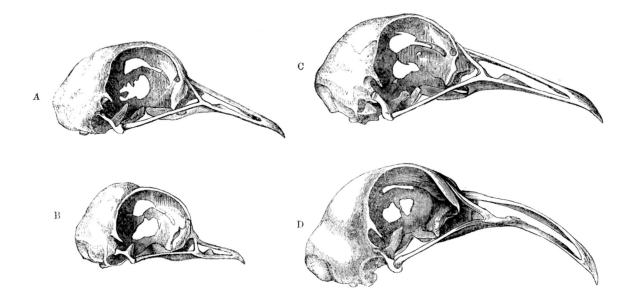

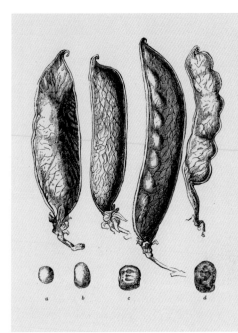

PANGENESIS

Darwin proposed what he called his "provisional hypothesis of pangenesis" as a theory of heredity, and it is a well known case in which Darwin was mistaken. The experiments of Gregor Mendel were barely known, and DNA was not discovered until long after Darwin's death. The word pangenesis was based on Greek roots, "pan" meaning "whole" or "encompassing", and "genesis" meaning "birth". Darwin speculated that each cell in an organism gave off tiny units of heredity, which he called "gemmules". These circulated freely throughout the system and thereby allowed each part to reproduce itself, or sometimes to be carried over generations without appearing. Although Darwin believed his theory was new, similar beliefs can be found in writings as far back as ancient Greece.

Drawings of the four most distinct kinds of pods and peas grown by Darwin.

are closely allied in structure, and we can at once understand the fact if they are likewise closely allied by descent. The succession of the many distinct species of the same genus throughout the long series of geological formations seems to have been unbroken or continuous. New species come in gradually one by one. Ancient and extinct forms of life often show combined or intermediate characters, like the words of a dead language with respect to its several offshoots or living tongues. All these and other such facts seemed to me to point to descent with modification as the method of production of new groups of species.

Darwin later recalled in his autobiography that the book "was begun … in the beginning of 1860, but was not published until the beginning of 1868. It is a big book, and cost me four years and two months' hard labour. It gives all my observations and an immense number of facts collected from various sources, about our domestic productions. In the second volume the causes and laws of variation, inheritance, &c., are discussed, as far as our present state of knowledge permits. Towards the end of the work I give my well abused hypothesis of Pangenesis."

From the beginning of his speculations about species in the 1830s the analogy with domesticated animals and plants struck him as the most powerful tool to understand how varieties and species arise in nature. The book examined many examples of domesticated plants and animals, especially pigeons, which Darwin had kept himself, as well as rabbits, fowls and ducks. He demonstrated the degree and nature of changes that domesticated species had undergone while under the control of humans, and argued that breeds of the same species, such as dogs and pigeons, were probably descended from a single wild ancestral species rather than many separate wild species.

One of the main points of the book was to demonstrate that "No part of the organisation escapes the tendency to vary." As far as could be determined, organisms were highly malleable and not just identical copies. For example, he showed that "In certain pigeons the shape of the lower jaw, the relative length of the tongue, the size of the nostrils and eyelids, the number and shape of the ribs, the form and size of the œsophagus, have all varied."

Right: From the
literature on breeding,
Darwin found that the
Golden Spangled Polish
Bantams had changed
since the 1730s because
breeders had chosen
to favour different
characteristics.

These small naturally
occurring differences or
variations were artificially
selected by humans to
improve their breeds in a
desired direction. Darwin made
the point that pigeon breeders, for
example, could not cause a bird to have
a slightly larger tail. Instead, they
selectively bred from individuals
with the longest tails, excluding
those with shorter tails. By
continuing this process over
generations, extremely different
breeds had been produced, so
different that a naturalist seeing
one in nature would classify
it as a different species. Darwin
used this point to show that since "we
have abundant evidence of the constant
occurrence under nature of slight individual
differences of the most diversified kinds … we
are led to conclude that species have generally
originated by the natural selection, not of
abrupt modifications, but of extremely
slight differences."

VARIATION IN NATURE

"If then organic beings in a state of nature vary even
in a slight degree, owing to changes in the surrounding
conditions, of which we have abundant geological
evidence, or from any other cause; if, in the long course
of ages, inheritable variations ever arise in any way
advantageous to any being under its excessively complex
and changing relations of life; … then the severe and
often-recurrent struggle for existence will determine that
those variations, however slight, which are favourable
shall be preserved or selected, and those which are
unfavourable shall be destroyed." Charles Darwin,
Variation, volume 1, 1868, page 5.

Dun Devonshire Pony, with shoulder, spinal, and leg stripes. Darwin argued
that the stripes were a vestige of an earlier ancestral species.

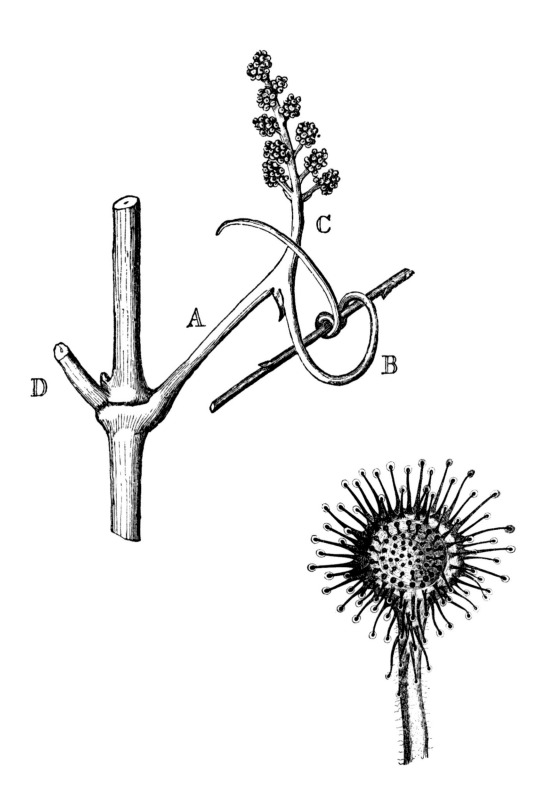

THE POWER OF PLANTS

As his health declined, Darwin worked more on plants because he could study them at home for short periods at a time. He became fascinated by the movement of plant tendrils, and eventually studied over 100 species of climbing plants, including twining plants, leaf-climbers and tendril-bearers.

His work on them stressed some familiar Darwinian themes –to make the invisible visible, and to explain the apparently inexplicable by revealing how tiny changes could build up to achieve great results.

When we see a climbing plant with its stem or tendrils wrapped around a stick we know that it must have gradually moved around the stick. But how? Because plants move so slowly many mysteries were involved with how they moved, such as whether light, gravity or touch affected their growth and motion. Darwin set to experimenting to find answers.

Long before time-lapse photography was possible, Darwin used several methods to reveal the motions of climbing plants. For example, he covered the plant to be studied with a glass dome or bell glass. He then traced the motion of the part of the plant he was observing onto the glass at regular intervals, until he had a series of lines, which could be transferred to paper, revealing how the plant had moved. This enabled Darwin to reconstruct what we can now see in a David Attenborough film such as *The Private Life of Plants.*

Darwin showed that many types of complex climbing plants had probably descended from earlier, simpler forms of climbing, such as wrapping the entire stem around an object rather than using specialized tendrils. He showed, for example, that "tendrils consist of various organs in a modified state, namely, leaves, flower-peduncles, branches, and perhaps stipules."

He first published his results in the *Journal of the Linnean Society* in 1865. This was later revised and published

Opposite, top: Flower-tendril of a grape vine showing its intermediate gradation between a true tendril, right.

Opposite bottom: Enlarged view of an insect-catching leaf of a sundew.

as a book by John Murray in 1875. Darwin concluded his work with the observation "It has often been vaguely asserted that plants are distinguished from animals by not having the power of movement. It should rather be said that plants acquire and display this power only when it is of some advantage to them; this being of comparatively rare occurrence, as they are affixed to the ground, and food is brought to them by the air and rain."

Darwin's book *Insectivorous Plants* (1875) contained meticulous studies of the adaptation of plants which live in nutrient-poor environments to extract nutriment from insects. He studied the sundew or *Drosera* in particular detail. The leaves of the sundew are covered with hundreds of filaments or tentacles tipped with glands surrounded by a sticky liquid. When an insect alights on one of these leaves it becomes trapped by the sticky droplets. The leaf of the plant then slowly curls inwards, enveloping the insect before digesting it. Darwin conducted experiments to discover what sorts of contacts and substances would activate the leaves. Willing to try almost anything, he applied substances such as hair, thread, milk, meat and glass, and found that animal chemicals excited action and not other substances. He discovered that the chemical substances produced by

the plant to digest insects were similar to the digestive juices of animals.

Stimuli that would waste the actions of the leaves, such as rain drops or sudden taps from a leaf or stick, caused no action. He investigated the precise actions of the cells of the tentacles and glands under his microscope in successive stages of action. Darwin suggested that natural selection could explain such extraordinary features, as there were ordinary plants with sticky glands which occasionally caught insects by accident, but did not digest them. Such plants "might thus be converted under favourable circumstances into a species capable of true digestion. It ceases, therefore, to be any great mystery how several genera of plants, in no way closely related together, have independently acquired this same power." The book also examined other sorts of carnivorous plants, such as the famous Venus flytrap.

In one of his last books, *The Power of Movement in Plants* (1880), Darwin argued that

Above: Diagram showing the motion of the upper internode of a pea plant between 8:46am and 9:15pm from *On the Movements and Habits of Climbing Plants* (1875).

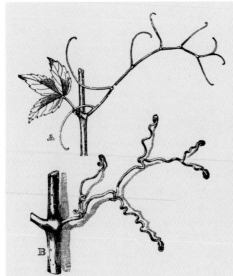

CLIMBING PLANTS

"I was led to take up this subject by reading a short paper by Asa Gray, (published in 1858), on the movements of the tendrils of a Cucurbitacean plant. He sent me seeds, and on raising some plants I was so much fascinated and perplexed by the revolving movements of the tendrils and stems, which movements are really very simple, though appearing at first very complex, that I procured various other kinds of Climbing Plants, and studied the whole subject. ... Some of the adaptations displayed by climbing plants are as beautiful as those by Orchids for ensuring cross-fertilisation." Charles Darwin, *Autobiography*, 1958, page 129.

Tendrils of Virginia Creeper, which Darwin showed are adapted for grasping flat surfaces rather than twigs.

A REMARKABLE DISCOVERY

"During subsequent years, whenever I had leisure, I pursued my experiments, and my book on *Insectivorous Plants* was published July 1875, – that is sixteen years after my first observations. The delay in this case, as with all my other books, has been a great advantage to me; for a man after a long interval can criticise his own work, almost as well as if it were that of another person. The fact that a plant should secrete, when properly excited, a fluid containing an acid and ferment, closely analogous to the digestive fluid of an animal, was certainly a remarkable discovery." Charles Darwin, *Autobiography*, 1958, page 132.

Title page of *Insectivorous Plants* (1875).

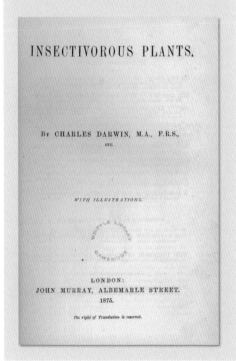

Right, top: White Bryony tendril attached to Grape Hyacinth. Darwin went outdoors during gales to observe the elastic hold of spiralled tendrils retaining their grip despite strong winds.

Right, middle: Leaf of an insect-catching Venus flytrap, which grows only in damp-poor soil in parts of North and South Carolina. Digested insects compensate for the lack of nutrients the plant can acquire through its roots.

the circular movements of growing stems had been adapted to permit many other kinds of motion:

It has now been shown that the following important classes of movement all arise from modified circumnutation, which is omnipresent whilst growth lasts, and after growth has ceased, whenever pulvini are present. These classes of movement consist of those due to epinasty and hyponasty, – those proper to climbing plants, commonly called revolving nutation, – the nyctitropic or sleep movements of leaves and cotyledons, – and the two immense classes of movement excited by light and gravitation. When we speak of modified circumnutation we mean that light, or the alternations of light and darkness, gravitation, slight pressure or other irritants, and certain innate or constitutional states of the plant, do not directly cause the movement; they merely lead to a temporary increase or diminution of those spontaneous changes in the turgescence of the cells which are already in progress.

THE EXPRESSION OF THE EMOTIONS

Darwin had collected notes on the origins of humans since the 1830s "with the determination not to publish, as I thought that I should thus only add to the prejudices against my views."

It seemed to Darwin sufficient to state at the end of *On the Origin of Species* "Light will be thrown on the origin of man and his history." After *On the Origin of Species* a few other authors, notably T H Huxley with his 1862 *Man's Place in Nature*, wrote about the evolutionary implications to mankind. Darwin thought it was time to give his own view and his work was published in two volumes in 1871 as *The Descent of Man, and Selection in Relation to Sex*.

An apparently odd feature of the work is that much of it is about sexual selection in other species, such as insects, fish, lizards, primates and especially birds. This was Darwin's second mechanism for explaining evolutionary change, though now regarded as a sub-type of natural selection. He explained sexual differences, such as male antlers, spurs on cocks or the peacock's tail as the result of differential success in males either competing against other males or being chosen by females and therefore leaving more offspring.

Darwin explained why sexual selection occupied so much of the work:

During many years it has seemed to me highly probable that sexual selection has played an important part in differentiating the races of man … When I came to apply this view to man, I found it indispensable to treat the whole subject in full detail. Consequently the second part of the present work, treating of sexual selection, has extended to an inordinate length, compared with the first part; but this could not be avoided.

The first part is what makes *The Descent of Man* so famous because it demonstrates with overwhelming evidence what humans are and what we come from. Darwin grouped the evidence into three kinds. The first of these included similarities between man and other

Opposite: Darwin argued that female birds could appreciate the nuances of the beautiful plumes of exhibiting males.

Right: "No other member of the whole class of mammals is coloured in so extraordinary a manner as the adult male mandrill (*Cynocephalus mormon*). The face at this age becomes of a fine blue, with the ridge and tip of the nose of the most brilliant red." *Descent*, volume 2, page 292.

MAN AND HIS ANIMAL RELATIONS

Darwin famously declared in his conclusion to *The Descent of Man*: "man is descended from a hairy quadruped, furnished with a tail and pointed ears, probably arboreal in its habits, and an inhabitant of the Old World. This creature, if its whole structure had been examined by a naturalist, would have been classed amongst the Quadrumana, as surely as would the common and still more ancient progenitor of the Old and New World monkeys. The Quadrumana and all the higher mammals are probably derived from an ancient marsupial animal, and this through a long line of diversified forms, either from some reptile-like or some amphibian-like creature, and this again from some fish-like animal. In the dim obscurity of the past we can see that the early progenitor of all the Vertebrata must have been an aquatic animal, provided with branchiæ, with the two sexes united in the same individual, and with the most important organs of the body (such as the brain and heart) imperfectly developed." Charles Darwin, *The Descent of Man*, volume 2, 1871, page 389.

Darwin argued that the manes and coloured skins of monkeys were sexual ornaments to attract females.

animals, such as in the bones in our skeletons, our muscles and organs etc. "On any other view the similarity of pattern between the hand of a man or monkey, the foot of a horse, the flipper of a seal, the wing of a bat, &c., is utterly inexplicable." He showed that humans are more similar to apes than apes are to any other living animals.

Second, he stressed the similarities in embryological development. Human embryos do not start out as very tiny mini adult humans which just get bigger, instead they go through a long developmental process in which they closely resemble the embryos of other animals. Finally, he discussed vestigial parts which no longer have a function but are a remnant from

ancestral forms, such as our rudimentary tail bones. Darwin also speculated that humans originally evolved in Africa.

Darwin had intended to discuss human expressions in *The Descent*, but the book became too long so he published a second work in 1872 – *The Expression of the Emotions in Man and Animals*. He was originally stirred to write on this subject by the assertion of an earlier writer who claimed that man is provided with special muscles just for expressing emotions. Darwin showed instead that human emotions and their expression were present to some degree in other animals. "It seemed probable that the habit of expressing our feelings by certain movements, though now rendered innate, had been in some

Opposite: Darwin's 1868 conjectural diagram about the relationship of primates, including man.

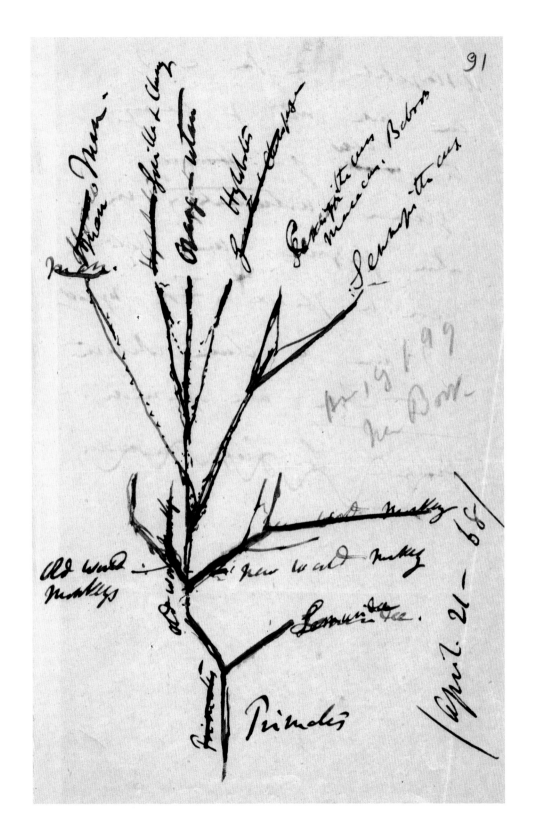

manner gradually acquired. But to discover how such habits had been acquired was perplexing in no small degree. The whole subject had to be viewed under a new aspect, and each expression demanded a rational explanation." He also showed that the main expressions were universal in all human races, which was additional evidence that all are descended from "a single parent-stock". Because of the difficulties of observing the fleeting expression of emotions Darwin used photographs in addition to the usual woodcuts, one of the first books to do so.

Right: Darwin showed that the facial muscles for expression are not a special human endowment as claimed by earlier writers.

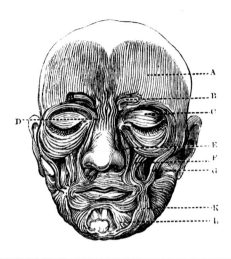

Opposite: An artist's pencil drawing of a dog similar to those published in *Expression* (1872). The upper one shows a "dog approaching another dog with hostile intentions" and the lower a dog "in a humble and affectionate frame of mind" – as Darwin captioned the published versions.

THE THREE PRINCIPLES OF EXPRESSION

Darwin argued that there were three general principles of expression.

1. "Serviceable actions become habitual in association with certain states of the mind, and are performed whether or not of service in each particular case".
2. "Antithesis. The habit of voluntarily performing opposite movements under opposite impulses has become firmly established in us by the practice of our whole lives. Hence, if certain actions have

been regularly performed, in accordance with our first principle, under a certain frame of mind, there will be a strong and involuntary tendency to the performance of directly opposite actions, whether or not these are of any use, under the excitement of an opposite frame of mind".
3. "The direct action of the nervous system … on the body, independently of the will and in part of habit".

Charles Darwin, *Expression*, 1872, pages 27, 348, 29 and 66.

According to Darwin we may imagine one of the figures on the left side to have just said, "'What do you mean by insulting me?' and the figure on the right side to answer, 'I really could not help it.' The helpless man unconsciously contracts the muscles of his forehead which are antagonistic to those that cause a frown, and thus raises his eyebrows; at the same time he relaxes the muscles about the mouth, so that the lower jaw drops." Charles Darwin, *Expression*, 1872, page 272.

THE SAGE OF DOWN AND THE STUDY OF WORMS

In his final years Darwin was seen by many as the world's greatest living scientist. His lifetime of important and often revolutionary scientific work had earned him many distinguished medals and prizes. He was honoured by scores of foreign scientific associations, and delegations of admirers begged for the opportunity to visit him.

Autograph hunters plagued him with requests, and with the advent of practical photography and cartes de visite, his portrait became familiar to everyone, since 1862 he began sporting his now-famous beard.

But Darwin was not interested in fame and was so excessively modest that he took a very different view of his own accomplishments. He concluded his autobiography with one of the most famous understatements ever written: "With such moderate abilities as I possess, it is truly surprising that thus I should have influenced to a considerable extent the beliefs of scientific men on some important points." To the end of his life Darwin continued to study plants, work at his microscope and publish scientific papers and letters in popular science magazines.

Darwin's last book was on earthworms, published in 1881 the year before his death. It was in many ways a very Darwinian work. Like so many of the others it was published decades after he first began speculating on the subject, and he once again discovered that small,

Opposite: Darwin in 1874 by Elliott and Fry. Colourized by the Natural History Museum, London.

Right: Darwin was honoured by universities and scientific institutions around the world with honorary memberships, degrees and medals.

(373)

APPENDIX IV.*

HONOURS, DEGREES, SOCIETIES, &c.

Order.—Prussian Order, 'Pour le Mérite.' 1867.
Office.—County Magistrate. 1857.
Degrees.—Cambridge { B.A. 1831 [1832].†
          { M.A. 1837.
          Hon. LL.D. 1877.
    Bonn . . Hon. Doctor in Medicine and Surgery. 1868.
    Breslau . Hon. Doctor in Medicine and Surgery. 1862.
    Leyden . Hon. M.D. 1875.
Societies.—London . Zoological. Corresp. Member. 1831.‡
          Entomological. 1833. Orig. Member.
          Geological. 1836. Wollaston Medal, 1859.
          Royal Geographical. 1838.
          Royal. 1839. Royal Society's Medal, 1853
              Copley Medal, 1864.
          Linnean. 1854.
          Ethnological. 1861.
          Medico-Chirurgical. Hon. Member. 1868.
          Baly Medal of the Royal College of Physicians, 1879.

Societies.—PROVINCIAL, COLONIAL AND INDIAN.
Royal Society of Edinburgh, 1865.
Royal Medical Society of Edinburgh, 1826. Hon. Member, 1861.
Royal Irish Academy. Hon. Member, 1866.

* The list has been compiled from the diplomas and letters in my father's possession, and is no doubt incomplete, as he seems to have lost or mislaid some of the papers received from foreign Societies. Where the name of a foreign Society (excluding those in the United States) is given in English, it is a translation of the Latin (or in one case Russian) of the original Diploma.
    † See vol. i. p. 163.
    ‡ He afterwards became a Fellow of the Society.

RELIGIOUS OPINIONS

Many people wrote to Darwin to ask about his religious opinions, something he chose not to publish. His son Francis recalled "He felt strongly that a man's religion is an essentially private matter, and one concerning himself alone." Many asked if one could believe in God and evolution. Darwin replied "It seems to me absurd to doubt that a man can be an ardent theist & an Evolutionist" because many people believed in both. "I may state that my judgment often fluctuates. … In my most extreme fluctuations I have never been an Atheist in the sense of denying the existence of a God. I think that generally (and more and more as I grow older), but not always, that an Agnostic would be the more correct description of my state of mind."

Francis Darwin as a young man. After his father's death he collected recollections and letters to produce *Life and Letters* (1887).

apparently trivial natural processes, ever present but unnoticed beneath our feet, completely change the surface of the land. According to Darwin, "Worms have played a more important part in the history of the world than most persons would at first suppose."

Darwin showed that the actions of worms were of great value to the soil by fertilizing it and aerating it with their burrows, which also allowed the soil to absorb more water and facilitated the growth of the roots of plants.

The book is full of his almost childlike enthusiasm to solve natural puzzles. Through careful measurements and experiments Darwin showed that "all the vegetable mould over the whole country has passed many times through, and will again pass many times through, the intestinal canals of worms." Earlier writers who had underestimated the ability of worms were criticized for their "inability to sum up the effects of a continually recurrent cause, which has often retarded the progress of science, as formerly in the case of geology, and more recently in that of the principle of evolution."

Darwin showed that the reason ancient ruins and artefacts are found underground is because they are undermined and buried by worms. Worms push up small piles of castings from their

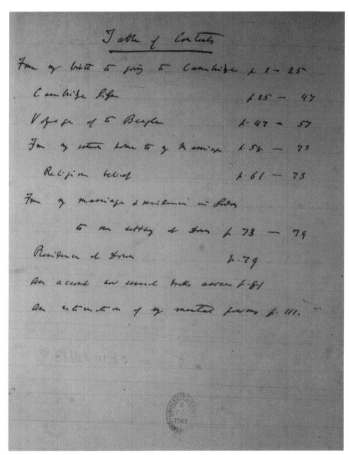

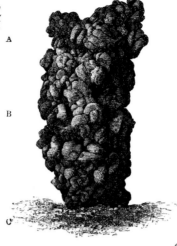

Left: For decades Darwin attended to and collected information on the piles or castings pushed up by earthworms.

Far left: Cross section of the earth in a field. The material in B has been pushed up by worms and thus buried the stones in C.

EXPERIMENTS WITH WORMS

Some of Darwin's most charming experiments were with worms. "Worms do not possess any sense of hearing. They took not the least notice of the shrill notes from a metal whistle, which was repeatedly sounded near them; nor did they of the deepest and loudest tones of a bassoon. They were indifferent to shouts, if care was taken that the breath did not strike them. When placed on a table close to the keys of a piano, which was played as loudly as possible, they remained perfectly quiet." He even tip-toed downstairs at night with a lantern to see if worms kept in flower pots indoors would react to light.

Diagram of the digestive tract inside an earthworm.

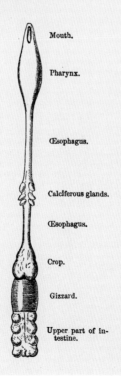

Mouth.

Pharynx.

Œsophagus.

Calciferous glands.

Œsophagus.

Crop.

Gizzard.

Upper part of intestine.

Opposite: Handwritten table of contents from Darwin's *Autobiography*, written for his family.

Above: According to Darwin's original caption this shows a "Transverse section across a large stone, which had lain on a grass-field for 35 years. A A, general level of the field. The underlying brick rubbish has not been represented. Scale ½ inch to 1 foot".

burrows. These, in effect, constantly push up the surface of the land via all the pieces small enough to pass through a worm. Darwin calculated this mass was as much as eight tons per acre a year. Larger objects thus slowly sink into a sea of ever-rising worm soil. Darwin measured this with many experiments including stones at Stonehenge and "a large flat stone" in his garden – now known as the worm stone.

"When we behold a wide, turf-covered expanse, we should remember that its smoothness, on which so much of its beauty depends, is mainly due to all the inequalities having been slowly levelled by worms. It is a marvellous reflection that the whole of the superficial mould over any such expanse has passed, and will again pass, every few years through the bodies of worms." The book was immensely popular, and initially it sold more copies than *On the Origin of Species* had.

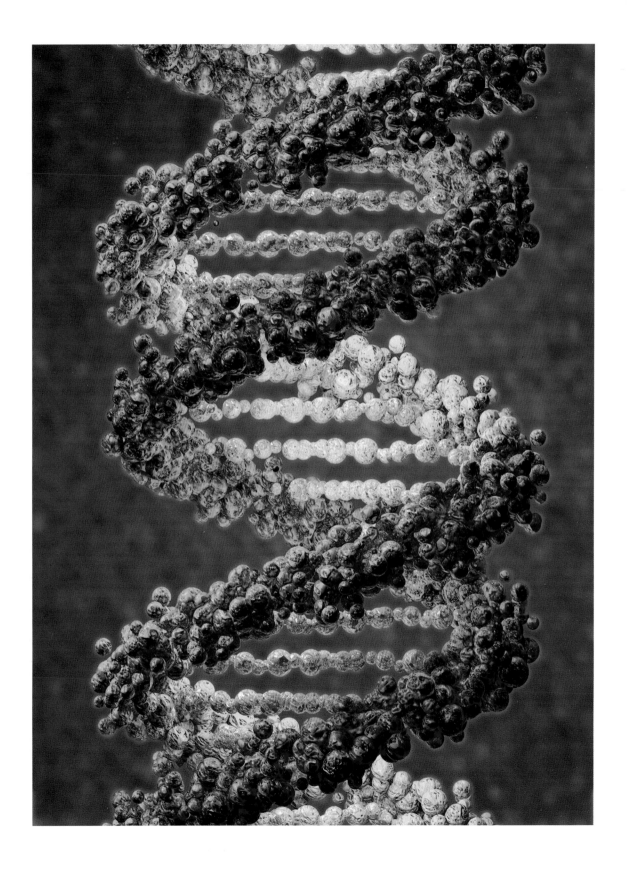

DARWIN'S DEATH AND LEGACY

Francis Darwin described his father's death as follows: "During the night of April 18th, about a quarter to twelve, he had a severe attack and passed into a faint, from which he was brought back to consciousness with great difficulty.

He seemed to recognize the approach of death, and said, 'I am not the least afraid to die.'... He died at about four o'clock on Wednesday, April 19th, 1882."

Since the early twentieth century a few legends have arisen, especially in America, claiming that Darwin recanted evolution or converted to Christianity on his deathbed. His family forcibly refuted these when they first appeared, but to this day they continue to circulate, generally among people who wish them to be true.

Although Darwin intended to be buried in St Mary's churchyard in Down, many senior figures in British science and the Church of England felt that their distinguished countryman should be interred in Westminster Abbey. It was soon arranged, and Darwin was

Opposite: DNA was not discovered until many years after Darwin's death. It revealed many of the mysteries of inheritance that puzzled Darwin.

Right: "Funeral of the late Charles Robert Darwin in Westminster Abbey" from *The Graphic*, 29 April 1882.

buried after a state funeral on 26 April 1882.

In the century and more since Darwin's death an amount of knowledge about the workings of living things and the history of the Earth has been uncovered which is without parallel in human history. Yet all of this work has confirmed and corroborated Darwin's essential points to a degree he could never have imagined. Further discoveries in the fossil record, the discovery of genetics and DNA and a host of other findings have made Darwin's theory of evolution as solid today as the theory of gravity. Darwin's theory makes sense of the whole natural world.

For example, DNA demonstrates with the greatest possible detail and precision the fact of the genealogical relationship of all living species. Plate tectonics has made sense of much of the once-mysterious distributions of plants and animals, such as the marsupials living only in America and Australia.

Countless transitional forms have been found in the fossil record which fill in some of the blanks between ancient ancestral groups

Above: *Ambulocetus* (walking whale), an extinct terrestrial ancestor of whales. One of the very many intermediate forms that have been found as Darwin predicted.

DARWIN'S OBITUARIES

Darwin's death was very widely reported and hundreds of obituaries were published in many different countries. They leave us in little doubt how highly Darwin was regarded by his contemporaries:

"The greatest naturalist of our time, and, perhaps, all time".
– *Daily Telegraph*

"Perhaps no student since man first began to speculate on the world which surrounds him ever attained ideas so far in advance of what had been deemed true, and saw these ideas find acceptance with his contemporaries." – *Daily News*

"the greatest scientific discoverer of his age and country".
– *New York Times*

"… he, more than any one man who ever lived, put the coping-stone upon the work of centuries, and impressed the whole coinage of thought with his own mint-mark." – *The Academy*

Darwin's death sparked hundreds of obituaries like this one printed in Magdeburg, Germany.

and modern groups such as *Ichthyostega*, an intermediate between fish and amphibians found first in Greenland in 1931. Such early tetrapods are the ancestors of all land-living vertebrates, in particular one with five finger-like bones in its fins (which is why we all have five fingers and toes). The recently discovered *Ambulocetus* the "walking whale" connects modern whales with mammalian crocodile-like animals. There are literally thousands of further examples of transitional forms connecting fish to amphibians, amphibians to early reptiles, reptiles to early mammals, reptiles to birds and non-human apes to humans. In one sense all species and fossils are transitional. Each is a single point on a long and continuous line of generations, halted only by complete extinction.

Of course human ancestors are the most exciting of these to many people. It is no longer the case, as it was in Darwin's day, "that connecting-links have not hitherto been discovered". Now many transitional forms have been found which connect primitive apes to modern humans, such as Peking Man (*Homo erectus*) and Lucy (*Australopithecus afarensis*).

A REVOLUTION IN HUMAN THOUGHT

Alfred Russel Wallace described Darwin as "the Philosopher who has wrought a greater revolution in human thought within a quarter of a century than any man of our time – or perhaps of any time … he has given us new conceptions of the world of life, and a theory which is itself a powerful instrument of research; has shown us how to combine into one consistent whole the facts accumulated by all the separate classes of workers, and has thereby revolutionized the whole study of nature". Alfred Russel Wallace, as quoted by W B Carpenter, *Charles Darwin: his life and work. Modern Review* 1882, 3: 500–24, page 523.

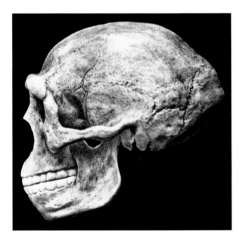

Above: Reconstructed skull of *Homo erectus*. As Darwin predicted, fossils of creatures between modern humans and earlier primates are being discovered.

Top right: Bronze medallion by Alphonse Legros 1881. Legros made a rough sketch of Darwin on the back of an envelope during a meeting at the Royal Society.

Right: Stone statue of Darwin by Horace Mountford (1905), outside the Old School, Shrewsbury.

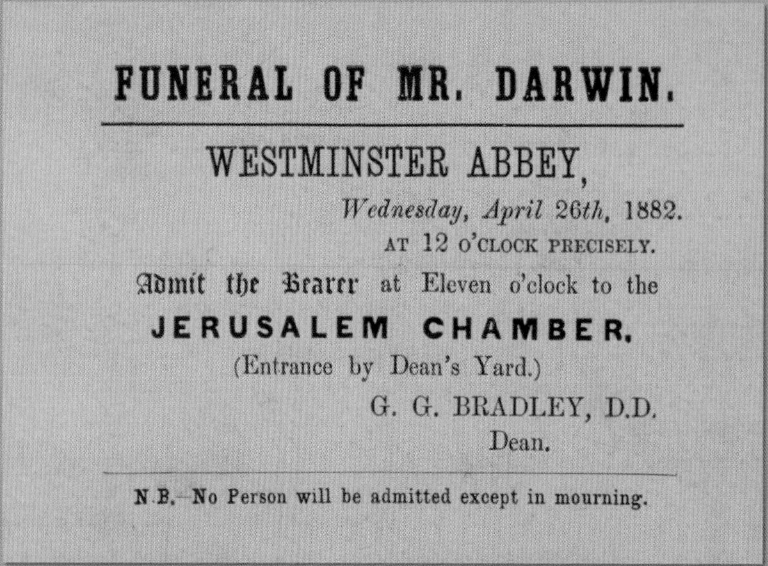

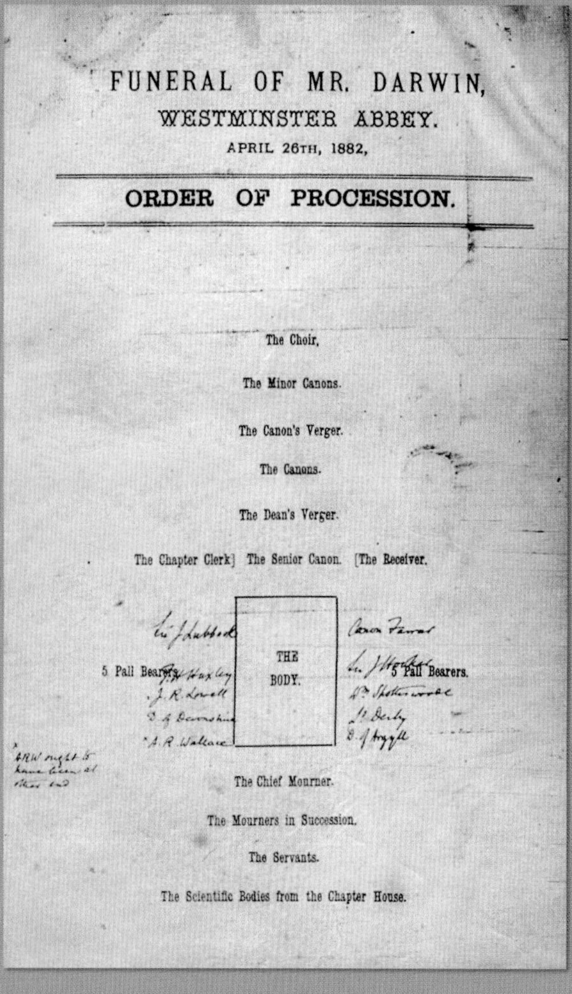

FUNERAL OF MR. DARWIN.

WESTMINSTER ABBEY,

Wednesday, April 26th, 1882.

AT 12 O'CLOCK PRECISELY.

Admit the Bearer at Eleven o'clock to the

JERUSALEM CHAMBER.

(Entrance by Dean's Yard.)

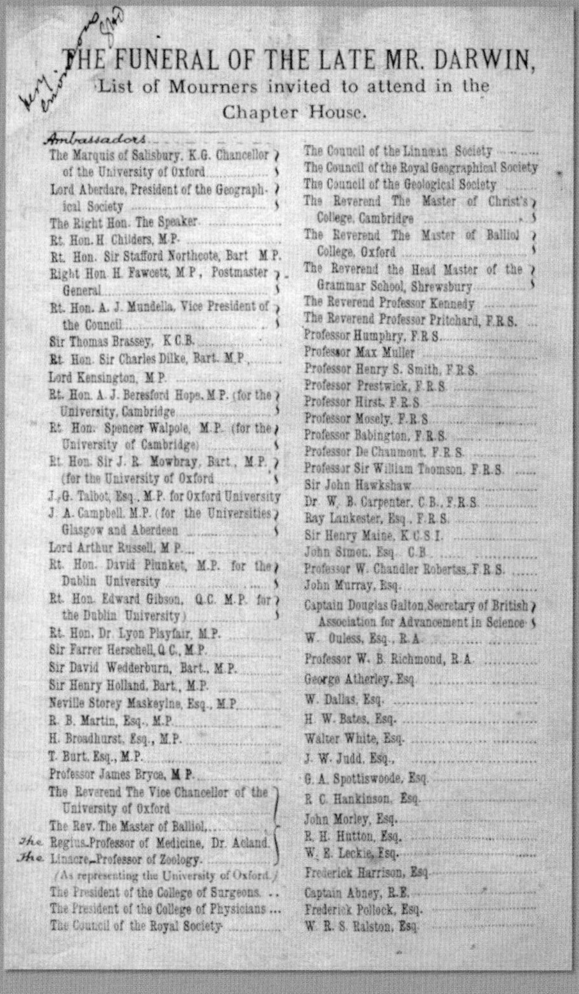

G. G. BRADLEY, D.D.

Dean.

N.B.—No Person will be admitted except in mourning.

Left: An admission card for Darwin's funeral. Attendees were assigned to different parts of the Abbey, this one is for the choir. Mourning dress was required.

Below and opposite: The printed order of procession for Darwin's funeral at Westminster Abbey on 26 April 1882 and a list of the most elite mourners selected to attend in the Chapter House.

Funeral Procession

Chief mourner

W. E. Darwin

	Mrs Wm Darwin
George Darwin	Mrs Litchfield
Miss Darwin	R. B. Litchfield
Francis Darwin	Mr Leonard Darwin
Horace Darwin	Mrs Horace Darwin
Mr Darwin of Elston	F. A. Darwin
Chas. Darwin	Reginald Darwin
Mr Vaughan Williams	Mrs Wedgwood
Charles Parker	Robt Parker
H. F. Bristowe	Francis Galton
Ernest Wedgwood	Hensleigh Wedgwood
Frances Wedgwood	Amy Wedgwood
× T. H. Farrer	~~Mrs Farrer~~
Godfrey Wedgwood	
Mr Ruck	Arthur Wedgwood
J. C. Hawkshaw	Mrs Hawkshaw
Geo. Allen	Henry Allen
Jackson	Parslow

× T. H. F. was too late to walk in the procession

SOURCES

Chapter 1
"we find no vestige" James Hutton *Theory of the Earth*, 1788.

Chapter 3
"curiously constructed" N Barlow ed., *The Autobiography of Charles Darwin 1809–1882. With the original omissions restored.* London: Collins, 1958, p. 22.

Chapter 4
"wretched microscope…" *Autobiography*, p. 50.
"During my second year" *Autobiography*, p. 52.
"a very pleasant and intelligent man" *Autobiography*, p. 51.
"[The chemistry] was the best part" *Autobiography*, p. 46.
"I also became" *Autobiography*, p.50.

Chapter 5
"the man who walks" Autobiography, p. 64.
"in old court" Letter to W E Darwin 15 October 1858, *The Correspondence of Charles Darwin*. vol 7, p. 170
"Upon the whole" *Autobiography*, p. 68. "one day, on tearing" *Autobiography*, p. 62.

Chapter 6
"If you can find" *Autobiography*, p. 71.
"a man of enlarged" J Wedgwood to R Darwin 31 August 1831, *Autobiography*, p. 230.
"Anxious that no opportunity" R. FitzRoy, *Narrative of the surveying voyages of His Majesty's Ships* Adventure *and* Beagle *between the years 1826 and 1836...* London: Henry Colburn, 1839, vol 2, p. 18.

Chapter 7
"for quoting the Bible" *Autobiography*, p. 85.
"showed me clearly" *Autobiography*, p. 77.

Chapter 8
"a man that knows every thing" FitzRoy, *Narrative*, p. 104.
"The Gauchos roared" Darwin, *Journal and remarks. 1832–1836.* London: Henry Colburn, 1839, p. 51.
"Ship not arrived" Darwin, "Falkland Maldonado (excursion) Rio Negro to Bahia Blanca" (2–5.1833; 8.1833). *Beagle* field notebook. EH1.14, [p. 129a].
"head, neck" Darwin, *Journal*, p. 108.

Chapter 9
"we may feel certain" Darwin ed., *The zoology of the voyage of H.M.S. Beagle. Fossil Mammalia. By Richard Owen.* London: Smith Elder, 1838, p. 5.

Chapter 10
"daily increases" R D Keynes ed., *Charles Darwin's Beagle diary.* Cambridge: University Press, 2001, p. 65.
"It was without exception…" Keynes, *Beagle diary*, p. 122.
"The skin is dirty copper" Keynes, *Beagle diary*, p. 122.

"a woman, who was suckling" Darwin, *Journal*, p. 235.
"For a moment, our position" FitzRoy, *Narrative*, p. 125.
"It was quite painful" Keynes, *Beagle diary*, p. 226.

Chapter 11
"An earthquake like" Keynes, *Beagle diary*, p. 292.
"The Earthquake & Volcano" Keynes, *Beagle diary*, p. 302.
"It is amusing to find" Keynes, *Beagle diary*, p. 340.
"Darwin's finches" P R Lowe, "The finches of the Galapagos in relation to Darwin's conception of species", *Ibis*, 1936, 310–21, p. 310.

Chapter 13
"saw several of those" Darwin, *Journal of researches*, 2d ed, p. 402.
"I had been lying" Keynes, *Beagle diary*, p. 402.
"Farewell Australia" Keynes, *Beagle diary*, p. 413.
"I received a letter" Autobiography, p. 81.
"the town is situated" Darwin *Journal of researches* 1839 p. 363
"When I recollect" N Barlow ed. 1963. Darwin's ornithological notes. *Bulletin of the British Museum (Natural History). Historical Series* vol 2, No. 7, pp. 201–278, p. 262.

Chapter 14
"after an absence", Darwin's personal "Journal" (1809–1881), Cambridge University Library (CUL) DAR158.1–76, p. 11.
"Freedom to go", Darwin, Memorandum on marriage. (7.1838) CUL DAR210.8.2.

Chapter 15
"One of the most" quoted in Darwin, *Journal of researches* London: John Murray, 1890, p. 5.
"When finding, as in this" *Journal*, p. 353.
"Unless we suppose" *Journal*, p. 399–400.
"the living sloths …" *Journal of researches*, 2d ed, p. 173.
"a most singular group…" Journal of researches, 2d ed, p. 173.
"Seeing this gradation" *Journal of researches*, 2d ed, p. 380.
"That some degree" Darwin, "On certain areas of elevation and subsidence in the Pacific and Indian oceans, as deduced from the study of coral formations" *Proceedings of the Geological Society of London* 2 (1837): 552–554, p. 544.
"On the connexion" Darwin, C R 1838. On the connexion of certain volcanic phænomena, and on the formation of mountain-chains and volcanos, as the effects of continental elevations. (Read 7 March) Proceedings of the Geological Society of London 2: 654–660.
"The most curious fact" Darwin, C R 1845. *Journal of researches into the natural history and geology of the countries visited during the voyage of H.M.S. Beagle round the world,*

under the Command of Capt. Fitz Roy, R N 2d edition. London: John Murray. pp. 379–80

Chapter 16
"their habits, ranges" Prospectus. In Darwin ed., *The zoology of the voyage of H.M.S. Beagle. Mammalia. By George R. Waterhouse.* London: Smith Elder and Co, 1838.
"it appears strange" *Mammalia*, p. 81.

Chapter 17
"In the early part of 1844" *Autobiography*, p. 116.

Chapter 18
"bang your bones" H E Litchfield ed. *Emma Darwin, A century of family letters, 1792–1896.* London: John Murray, 1915, vol 2, p. 221.

Chapter 19
"where does he do" F Darwin & A C Seward eds., *More letters of Charles Darwin.* London: John Murray, 1903, vol 1, p. 38.
"beloved" Darwin to R Owen, 26 March 1848, CCD4:126.
"confounded" Darwin to J D Hooker, 13 June, 1850 CCD4:343.
"I hate a Barnacle" Darwin to W D Fox 24, October 1852, CCD5:99.
"began sorting notes" CUL DAR158.1–76, p. 32.

Chapter 20
"to adapt & alter" Notebook B CUL DAR121, p. 4.
"species are not" Darwin to J D Hooker, 11 January 1844, Burkhardt F H et al eds. 1985–. *The correspondence of Charles Darwin.* Cambridge: University Press, vol 3, p. 2. (Henceforth CCD)
"When I saw your bundle" Darwin to O Salvin 12 October (1871) Calendar number 8005a private collection.
"In October" *Autobiography*, p. 120.

Chapter 21
"the theory of descent" Darwin, *On the origin of species by means of natural selection, or the preservation of favoured races in the struggle for life.* London: John Murray, 1859, p. 343.
"Light will be thrown" *Origin*, p. 488.
"no doubt the chief" *Autobiography*, p. 122.
"Hardly any" Darwin, C R 1866. *On the origin of species by means of natural selection, or the preservation of favoured races in the struggle for life.* London: John Murray. 4th edition. p. 367.

Chapter 22
"deserted the true", A Sedgwick, "Objections to Mr Darwin's theory of the origin of species", *Spectator* 1860 (24 March): 285–6, p. 286.
"I have read your" Sedgwick to Darwin 24 November 1859 CCD7:396.
"I have heard" Darwin to C Lyell, 10 December 1859, CCD7:427.
"Your leading idea" H C Watson to Darwin 21 November 1859, CCD7:385.

"The history of", W B Carpenter, "Darwin on the Origin of Species", *National Review* 1860, 10: 188-214, p. 214.

Chapter 24
"was begun…in the" *Autobiography*, p. 130.
"No part of the …" Darwin, The variation of animals and plants under domestication. London: John Murray, 1868, vol 2, p. 408.
"we have abundant" Variation vol 2, p. 414.
"To exhume" Darwin, C R 1868. *The variation of animals and plants under domestication.* London: John Murray. First edition, first issue. Volume 1. pp. 10–11

Chapter 25
"tendrils consist of" Darwin, *The movements and habits of climbing plants.* 2d ed. London: John Murray, 1875, p. 193.
"It has often been" *Climbing plants.* 2nd ed., p. 206.
"might thus be converted" Darwin, *Insectivorous Plants.* London: John Murray, 1875, p. 363.
"It has now been shown" Darwin, C R 1880. *The power of movement in plants.* London: John Murray. p. 569.

Chapter 26
"with the determination" *Descent* vol 1, p. 1.
"Light will be thrown" *Origin* p. 488.
"During many years" *Descent* vol 1, p. 5.
"On any other view" *Descent* vol 1, p. 31.
"It seemed probable" Darwin, *The expression of the emotions in man and animals.* London: John Murray, 1872, p. 19.

Chapter 27
"With such moderate" *Autobiography*, p. 145.
"Worms have played" Darwin, *The formation of vegetable mould, through the action of worms, with observations on their habits.* London: John Murray, 1881, p. 305.
"all the vegetable mould" *Worms*, p. 4.
"inability to sum up" *Worms*, p. 6.
"a large flat stone" *Worms*, p. 119.
"When we behold" *Worms*, p. 313.
"Worms do not possess" *Worms*, p. 26.
"He felt strongly" F Darwin ed. T*he life and letters of Charles Darwin.* London: John Murray, 1887, vol 1, p. 304.
"It seems to me absurd…" Darwin to J Fordyce 7 May 1879 CUL DAR139.12.12.

Chapter 28
"During the night" *Life and letters* vol 3, p. 358.
"that connecting-links" *Descent* vol 1, p. 185.
"the Philosopher who" W B Carpenter, Charles Darwin: his life and work. *Modern Review* 1882, 3: 500–24, p. 523.

INDEX

(page numbers in *italic* refer to illustrations and captions)

CREDITS

The publishers would like to thank the following sources for their kind permission to reproduce the pictures in the book.

Key: t = top, b = bottom, l = left, r = right and c = centre

Academy of Natural Sciences of Philadelphia: 111, 142
Alamy: Amoret Tanner: 137t
Peter Aprahamian: 78b
Bridgeman Images: Bible Society, London, UK: 6; /Bibliotheque Nationale, Paris, France: 5; /Bibliotheque Sainte-Genevieve, Paris, France: 45; / British Museum: 64, 65t; /Edinburgh University Library, Scotland: 20t; /Musee de la Ville de Paris, Musee Carnavalet, Paris, France: 125; / Private Collection: 8b, 20b, 40-41, 46tr; /Rafael Valls Gallery, London, UK: 42; /Royal Geographical Society, London, UK: 66
By permission of the Syndics of Cambridge University Library: 21br (MS. DAR.5;A.42), 22-25 (MS.DAR.5;A.06)
Cambridge University Library: 92, 97, 98, 101t, 101b, 145, 150, 151b
Christ's College Library: 28b, 113
English Heritage Photo Library: 16b, 43t, 72-73, 76t, 76b, 99, 100t, 101b, 116
FLPA: B.Borrell Casal: 41t
Getty Images: 46, 110, 141c, 155br; /Archive Photos: 12b; /Bettmann: 7, 12t, 28t, 69b, 155l; /CORBIS/Corbis via Getty Images: 8b, 29bl, 94t; /Michael Dunning: 152; /Hulton-Deutsch Collection/CORBIS/Corbis via Getty Images: 9b, 126; /Kean Collection/Hulton Archive: 133; /Colin Keates: 53tl, 53b; /Michael Maslan/Corbis/VCG via Getty Images: 65b; /Popperfoto: 21bl
Mary Evans Picture Library: 17c, 18-19, 26-7, 46b, 68t, 72, 153
By kind permission of the Master and Fellows of Christ's College, Christ's College Library, Cambridge: 30 (T1830-35), 31-33, 44 (MS.DAR.44;13), 49 (MS.DAR.44;29), 80-81 (MS.DAR.5;A.06.7), 86-87 (MS.DAR.210.08.02),

102-103 (MS.DAR.158), 114 (MS.DAR.158), 115 (MS.DAR.121), 124 (MS.DAR.140), 147 (MS.DAR.53.1;164), 156t (MS.DAR.A203), 156b-157 (MS.DAR.215)
Dr Milo Keynes: 16t
Harold Mernick: 78t
Museum of Earth Sciences, Cambridge: 13t
Natural History Museum: 1, 9tl, 9tr, 10, 11, 13bl, 13br, 36t, 48b, 52, 53t, 82, 108, 109, 112, 117, 127, 129, 148
National Maritime Museum, Greenwich, London: 34
Photoshot: NHPA/Joe Blossom: 134
REX: Eileen Tweedy/Shutterstock: 119tl
Brian Rosen: 95t
Royal Geographical Society: 36b
Science Photo Library: 154t
Science & Society: 21t
Shrewsbury Museum: 14, 15
Star Ledger/Will Perlman: 79b
Topfoto: 17b, 37t, 48t, 57b, 100b, 128, 155t
Wellcome Trust Library: 29br
Reproduced with permission from John van Wyhe ed, The Complete Work of Charles Darwin Online (http://darwin-online.org.uk): 2, 27, 29t, 37b, 38-39, 43c, 43b, 50t, 50b, 53tr, 56t, 57t, 58-59, 60-62, 63, 68b, 68t, 69t, 70tl, 70bl, 70br, 74-75, 83, 84-85, 88-91, 93, 94-95, 96, 104-107, 118, 119b, 119r, 120-123, 130-132, 135, 136, 137b, 138-139, 140, 141b, 143, 144, 146, 149, 151

Every effort has been made to acknowledge correctly and contact the source and/or copyright holder of each picture, and Carlton Books apologizes for any unintentional errors or omissions, which will be corrected in future editions of this book.